Masters of the Structural Aesthetic

Derek Thomas

Masters of the Structural Aesthetic

Springer

Derek Thomas
Cape Town, Western Cape
South Africa

ISBN 978-981-10-5444-0 ISBN 978-981-10-5445-7 (eBook)
DOI 10.1007/978-981-10-5445-7

Library of Congress Control Number: 2017946476

Printed on acid-free paper

This Springer imprint is published by Springer Nature
The registered company is Springer Nature Singapore Pte Ltd.
The registered company address is: 152 Beach Road, #21-01/04 Gateway East, Singapore 189721, Singapore

Preface

At the end of my final year in architecture, I had the good fortune to be awarded a travel scholarship to any place of my choice. The focus of the study had to contribute to the understanding of architecture as being the art of philosophical embodiment and of intuitive thought. Efficient buildings are often without architectural value, just as an initially pleasing building can be inefficient and quite possibly impractical. On my study tour to Italy, I immersed myself in finding out more about Pier Luigi Nervi who had managed through his intuitive skills and architectural sensibility to mix just the right proportion of both, but more significantly, to what extent he was operating in sync with Nature.

While exploring the work of Professor Nervi in more depth, I researched his new structural techniques and its potentialities. I realised at the time that Nervi was becoming an icon of the twentieth century for his achievements in both structural and constructional engineering, as well as in architecture. This view was reinforced by Ernesto Rogers (1957) about Nervi as expressed in *A Portrait of Pier Luigi Nervi* (quoting Leon Battista Alberti): 'Him I call Architect, who, by a sure and wonderful Art and Method, is able, both in thought and invention, to devise, and with execution, to complete all those Works, which, by means of Movement and great Weights, and amassment of Bodies, can, with the greatest Beauty, be adapted to the uses of Mankind.' Of all the commentators on Nervi's approach to design, starting with 'a basis of scientific certainty, but transcending through intuition' and his 'disciplined and precise mind', Rogers' clarity of observation stands out.

As though to confirm how elusive, even subjective, a unified evaluation of the structural aesthetic can be, even amongst those recognised as authorities, there have been discordant views. In *Ove Arup: Master Builder of the Twentieth Century,* Arup stated that Pier Luigi Nervi had said that 'the outward appearance of a building cannot and must not be anything but a visual expression of an efficient structural and constructional reality'(Jones 2006). Arup, however, initially disagreed that this maxim could be the 'main criterion of architectural merit'. It did not fit many obvious famous works (he stated) such as Le Corbusier's Ronchamp Chapel or several buildings by Mies van der Rohe or Oscar Niemeyer, and it overlooked his long-held belief that size and scale are always relevant to criteria of merit. 'Nervi's own obvious interest in form is not always the visual expression of structural and constructional reality', Arup observed in 1955, and he went further: 'Formal and spatial relationships on the one hand and structural rightness on the other can conflict: the idea that correct structural and the best possible aesthetic solutions are one and the same thing must, I'm afraid, be abandoned together with the older philosopher's dream about the harmony and ultimate identity of truth, goodness, justice and beauty'.

Due to the rising 'flamboyant rhetoric and incontinent generalisations [that] easily attracted indefensible attention... [Arup] who had denounced the idea that correct structural and the best possible aesthetic solutions are one and the same thing was forced to reiterate one year later: 'I am not really attacking Nervi—he is on the side of angels in this matter—I am defending him against those who use his words to defeat his spirit'. He added that building 'is an art within a limited framework within which certain rules must be obeyed. Nervi's kind of logical build up in a scheme is to my mind the essence of good engineering and good architecture alike. Architects forget about the more sordid aspects of life because they—quite legitimately—get involved in the aesthetic aspects of their work' (Jones 2006). This turnaround can be construed as an endorsement that indeed structure and the aesthetic are indispensable allies in the design of the best in architecture, as evident in the works of the ten master architects presented in this narrative.

This alternative evaluation of Nervi's work applies equally to the role of architects and the products of their thought and invention down the Ages and is the starting point for this narrative which explores ten masters of the structural aesthetic and their architectural legacies.

The history of architecture acknowledges the legendary role of architects who have led the field and through their example have brought about significant paradigm shifts. Architecture, like art, reflects the changing cultural mood of the times and is marked by movements, for example, the Renaissance, or Modernism and even the avant-garde. According to Iversen (1993), Alois Riegl claims that 'the change in the artistic forms and aesthetics are pretty much related with the change of the way people perceive the world.'

The role of the structural engineer is paramount in realising the designs of the architect. 'The aesthetics of a well designed structure is indeed inherent in its very existence and therefore, when visually expressed clearly and honestly makes its own aesthetic contribution' (Khan 1980).

More in-depth scientific knowledge of the nature and properties of materials and advanced technology have given to the modern workman immense advantages over his medieval brother-craftsman. 'The efforts of primitive man in the construction of dwellings must have been at first determined solely by his physical wants. Only after these had been provided for, and materials amassed on which his imagination might exercise itself, would he begin to plan and erect structures, possessing not only utility, but also grandeur and beauty' (Statham 1911). Computer-aided technology has brought about significant changes in the details of the trade, or the art of building, although stones and bricks used historically and concrete, and steel now, still form the more solid parts of all edifices.

The essential focus of this book is the structural aesthetics of ten seminal works of architecture and those architects and engineers responsible for them. This requires a holistic rather than a simplistic evaluation from a wider perspective that should take into account innovative digital design aids, structural innovation and advanced construction technologies.

A question arises as to whether structural honesty and integrity impose opportunity or constraint on the builders since through their limitations, materials as such have encouraged design ingenuity related to their particular properties and structural potentialities. History shows that there are numerous iconic building forms in basic materials, such as mud or clay, before concrete and steel freed the imaginations of the architect and structural engineer to design buildings that demonstrate the intrinsic potential of either material with honesty and integrity.

The study of geometry related to shape and form, space and motion from a visual standpoint, has been the subject of much interdisciplinary research into the fundamental problems of neuroscience (Sangari et al. 2011). So too have the ideas from the biological sciences affected architectural thought and practice where Nature is presented as the main source, currently known as biomimicry. In the sphere of genetic preference, there is strong evidence that evolution is not restricted to biology as we associate it with Charles Darwin, but can be observed in humankind's cultural selection. According to observers such as Richard Dawkins, a reality in cultural proclivity exists that relates to ideas and theories, elements that he calls 'memes'. This hypothesis cannot be summarily brushed aside in determining differing selectiveness between cultures in the Arts and has a bearing on cognition and thus on what constitutes distinct preferences in design (Distin 2005). Exploring the psychological argument further, art historian Gombrich argues 'there may be a transcendent aspect of art, a realm of values that great works of art incorporate, ultimately one's response to these values is based on conditioned taste... and not on the application of methods or rules.' He concludes that 'aesthetics, as rules or methods for discussion of aesthetic quality, cannot capture the values of art.' (Richmond 1994).

On the role of cladding of buildings, 'Some theorists denounce decoration of the underlying structure as architecturally improper, as evidenced in buildings that rely on the external treatment applied to the surface, when in fact architectural purity is of a much higher order' (Sparshott 1994). Others observe that the experiential aspect of appreciating architecture is as important as the architecture itself, questioning why certain architectural experiences stay with us whereas others vanish (Danielsson 2011).

Represented in this book on the aesthetic of the structure in architecture, a cross section of leaders in the field of the structural aesthetic from ancient times to present day are identified as having met the essential determinants and philosophic observations provided by recognised commentators. The selected protagonists are: The Roman master builders (2BCE–4CE), the Gothic architects (1300–1500), Filippo Brunelleschi (1377–1446), Antoni Gaudi (1852–1926), Le Corbusier (1887–1965), Pier Luigi Nervi (1891–1979), Oscar Niemeyer (1907–2012), Jörn Utzon (1918–2008), Frei Otto (1925–2015) and Zaha Hadid (1950–2016).

> Structural architecture leads to that synthesis of static-aesthetic sensitivity, technical knowledge and mastery of execution which produced masterpieces of the past.
> Pier Luigi Nervi, 1958

Keywords

Structural aesthetic; Biomimicry; Fractal design; Thematic transformation; Parametric design; Sense of place; Cultural proclivity; Cognitive associations; Symbolic design; Visual harmony

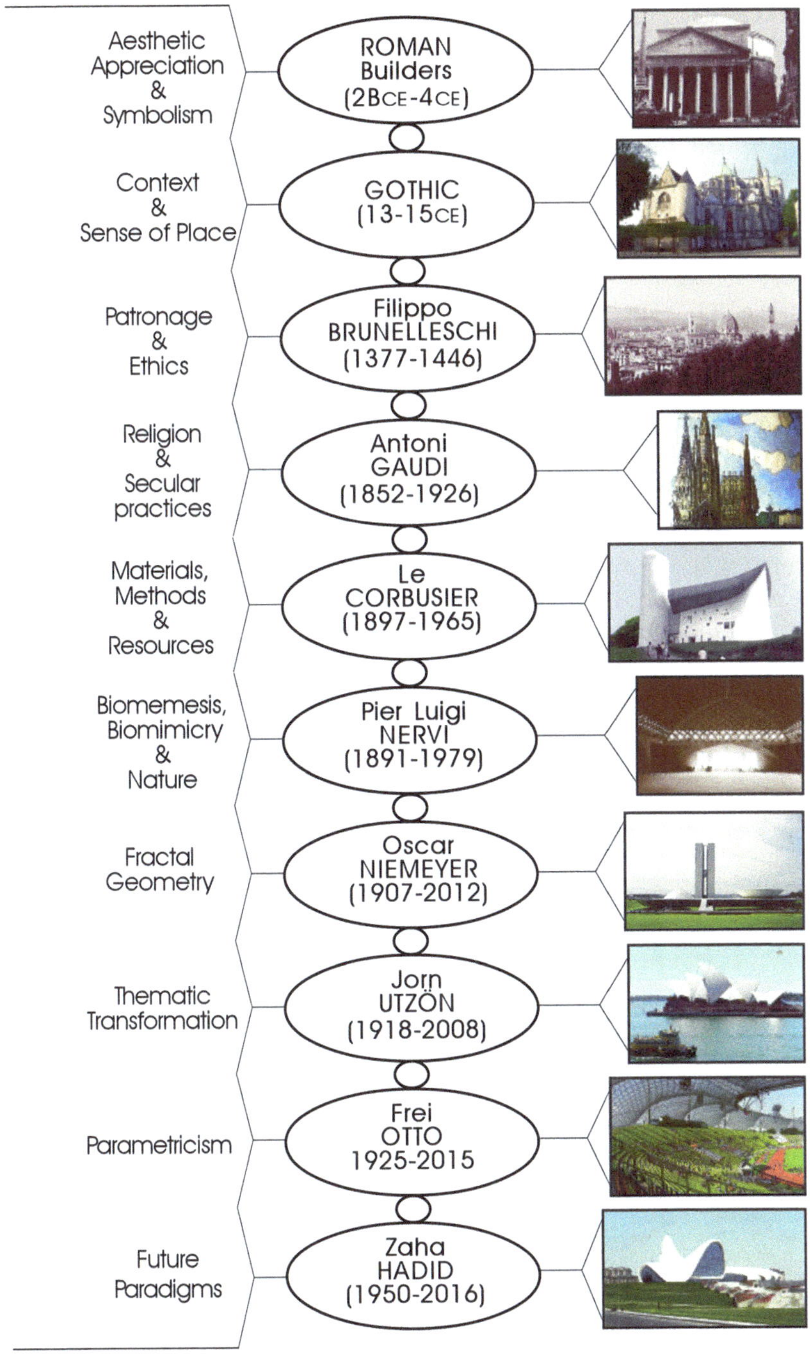

Fig. 1 Elemental determinants

References

Danielsson, C. B. (2011). *Experiencing architecture – Exploring the soul of the eye* (pp. 95–101). Stockholm: The Royal Institute of Technology (KTH).

Distin, K. (2005). *The selfish meme: A critical assessment* (pp. 2–5). Cambridge: Cambridge University Press.

Iversen, M. (1993). *Alois Riegl: Art history and theory* (p. 13). Cambridge: The MIT Press.

Jones, P. (2006). *Ove Arup: Master builder of the 20th century* (pp. 195–196). New Haven, CT: Yale University Press.

Khan, F. R. (1980). Structural aesthetics in architecture and its social and technological relevance. In Article in *Zeitschrift: IABSE congress report AIPC=IVBH, Kongressbericht, Nov*. (p. 136).

Richmond, S. (1994). Gombrich's critique of aesthetics. In *Aesthetic criteria and the philosophies of science of Popper & Polyani* (11th ed.). Amsterdam: Rodopi B.V.

Rogers, E. N. (1957). *The works of Pier Luigi Nervi* (vi, ix, x, xi) (Preface by Pier Luigi).

Sangari, A., Mirkia, H., & Assadi, A. H. (2011). *Perception of motion and architectural form: Computational relationships between optical flow and perspective*. Wisconsin: University of Wisconsin.

Sparshott, F. (1994). The aesthetics of architecture and the politics of space. In M. H. Mitias (Ed.), *Philosophy of architecture* (p. 13). Amsterdam: Rodopi.

Statham, H. H. (1911). Architecture and dome. In *Encyclopaedia Britannica* (Vols. 2 & 4).

Acknowledgements

All sources that were researched to inform this discourse have been recorded in the reference listings appearing after each chapter. The majority of the images appearing in the Figures are from the author's album. The following are acknowledged with much appreciation for their permission to use their images: Anton Luger, for the Sydney Opera House images; Tobias Alt, for the Munich Stadium images; Judy Maguire, for the image of the Djenne Mosque in Mali; Marcelo Jorge Vieira, for the images of the National Congress Building, Brasilia; and Hugh Fraser for the images of the Heydar Aliyev Center, Baku, Azerbaijan. The RIBA's permission to use the full record of Pier Luigi Nervi's acceptance speech is acknowledged.

Contents

Chapter 1
Introduction

Abstract In this chapter, the aim is to give substance to the premise that the 'structural aesthetic' is valid as a measure of architecture of worth, an aspect that is largely overlooked in the wider context. To this end a virtual matrix of criteria, backed by the philosophical research of recognised commentators, becomes a filter to guide the selection of the most deserving ten 'masters' and their landmark buildings. As contributing factors, cognitive associations and cultural preference as explored by neuroscience are fundamental to architectural appreciation. The influences of manual methods to the empowering technical advances are seen as integral in the wider perspective. These parameters coupled with the reliance on Nature of master builders from the past and contemporary architects as an inspiration for their architectural forms at all periods of architectural history are compelling in the design equation that has assumed many formulations, from two-dimensional geometry of earlier times to three-dimensional parametricism of recent works.

From the field of architecture there are many architects who remain acknowledged as 'masters of architectural design', and who justify the highest honours as singular beacons for their individual contributions to the history of the noble art of architecture. However, their accreditation, essentially as 'masters of architectural design', relates more to their contributions to the language expressed in architectural design, form and style and not primarily to that of 'the structural aesthetic' per se.

Amongst those who contributed significantly to how architecture presented itself was Mies van der Rohe whose 'anti-expressionist, anti-decorative position' is clear in his scorn for 'a senseless and trivial chaos of forms'. In search of a rational method, he called for the derivation of building from the nature of the problem, setting a course he would pursue for the rest of his life. And though van der Rohe emphasised 'structural character', he rationalised and was capitivated by the application of glass—which forced the direction of architecture into new ways by celebrating the shimmering expanse of glass wall cladding without the predominance of building's structure. The planal surface and pure materials became indispensable elements of design that were used internationally in a style that defined new design conception.

D. Thomas, *Masters of the Structural Aesthetic*, DOI 10.1007/978-981-10-5445-7_1

The Modern movement generally was characterised by an attack on adornment, yet, except for the few, the principle that the structure itself as the main force in the attainment of aesthetic excellence was not always embraced. Rather, as stated by Adolph Loos, the Austrian architect, the Movement tended to philosophise about the evolution of culture being "synonymous with the removal of ornament" and "Freedom from ornament is a sign of spiritual strength" (Loos 1971).

In the post-Modern era, Frank Gehry has captured a sense of motion—'frozen motion'—in his architecture. His twisting and turning metal forms some say reflect the influence of film, which simulates motion through a sequence of still images. Yet the simulated movement in this architecture seems to have no direction, and to be going nowhere. As architectural critic Michael Sorkin has observed, where early modern architecture simulated movement that was directional and where the arms of a Bauhaus building by Walter Gropius and planes of Mies Van der Rohe's Barcelona Pavilion reached out into the landscape.... Gehry's rotating masses and tipping facades show "the stationary dynamism of a flickering flame or cinematic image" (Gartman 2009).

Further afield, China was awarded to host the Olympic games in 2008. In a competitive field of architects from all over the world the winning design was that of Swiss architects, Herzog and de Meuron, for what has been considered to be a marvel of structural engineering. Seating 91,000 spectators, and built to withstand earthquakes of up to 8.00 on the Richter Scale, the 'Bird's Nest'—associated with a bird's nest that is an harmonious reference to Nature, or symbolises the crackle glaze on a Chinese ceramic vase, or with the traditional dish eaten on special occasions—is a unique example of exoskeletonism with its interwoven box section lattice construction. The execution of the design demanded unconventional methods requiring workers to crawl over the superstructure like an army of ants. Arup Sport provided the structural expertise with local partner China Architecture Design & Research Group. It would not do justice to the collective expertise to single out any of the disciplines involved with its conception and construction, however, the collaboration of Chinese artist Ai Weiwei was sought in this defining architectural concept for modern China. WeiWei eventually withdrew from the project. After stating that "the stadium has become a shining symbol of China's march towards modernism" he pronounced it pretentious and "a kind of fake image". In so doing he did not entirely endorse its eligibility as an example of structural honesty. Apart from the stadium's original deployment of the exoskeleton principle, arguably it structural fame relies more in the extraordinary and complex execution of the structure itself.

To refine the focus of the discourse about the 'structural aesthetic', Sparshott (1994) raises the issue as to whether it is valid to single out the structure where it predominates, and questions whether the structural aesthetic depends on the 'whole' in the evaluation of the architecture itself. "The primacy of the structural has often led aestheticians of architecture to argue that the most beautiful buildings are those in which elegant, economical, and imaginative solutions to problems in construction are most clearly or expressively visible. It follows from this that a building in which the forces apparently operative in its strength and stability are not those which are really at work, is a dishonest building to be condemned accordingly " (Sparshott 1994).

> Buildings are like a delicately woven fabric, a system of nerves made from structural elements purged of all frills so as to seem always in vibration, an equilibrium of tensions and of restrained, disciplined movement.
> Rogers (1957)

Pier Luigi Nervi's work provides proof that architectural expression and originality are able to combine the difficult and complicated workings of modern practice, and to be efficient, practical, economical, through the aesthetic of the structure. For a complete understanding and appraisal of his buildings, each of these characteristics of Nervi's work are interdependent but also places him amongst foremost thinkers from collaborative fields that today are widely separated through specialisation. "Constructional complications, or designs that require structural acrobatics, are always a sign of a false structural conception—even to the untrained eye of the observer. A consideration of this danger brings us to what, to my mind, is now the most important problem in architecture: training the Architect of Tomorrow" (Nervi 1955).

The work and structural methods of Nervi open the vista to the work of others, similarly imbued with the quest for the perfect equilibrium between the honesty of the structure in the attainment of fine architecture. Where Nervi's philosophies, his practical and artistic sensibilities make him one of the forerunners in his field, he was subliminally in tune with other disciplines such as mathematicians, physicists, biologists who were connecting through a new science—that of non-linear geometry that spawned its own language, "an elegant shop talk of fractals and bifurcation" (Gleick 1997).

Apart from Nervi, other modern architects who are celebrated in this book for their landmark structures that have stretched the boundaries of conventional practice are: Le Corbusier (1887–1965), Oscar Niemeyer (1907–2012), Frei Otto (1925–2015), Jörn Utzon (1918–2008), and Zaha Hadid (1950–2016). Casting the net even wider retrospectively, was the work of Filippo Brunelleschi (1377–1446), whose structures expressed the early Renaissance iconically, and Antoni Gaudi (1852–1926) who searched for "the presence of 'beauty' in force lines and funicular polygonals, with the consequence of preferring helicoids and parabolas with respect to circles and squares" (Lorenzi and Mauro 2010). Even much earlier, the Roman builders (2BCE–4CE) broke new ground and developed the use of structural principles such as the arch and dome. Later the Gothic master craftsmen (1300–1500) took structural aesthetics to unimaginable heights and extended the application of static laws and "threw the rein on the neck of experiment, utilising stone to its ultimate capacity" (Fletcher 1948).

To advance the understanding of what influences predetermine 'the structural aesthetic' a distillation of relevant criteria is fundamental to aid the process. In

Chap. 2, ‘elemental determinants’ are researched from a plethora of commentators and literature. To define aesthetics and its cognitive associations is a starting point, where issues such as subjectivity are counterpoised with those of scientific measure through neuroscience. Form and shape, style and cultural influences are but a sampling to ignite a greater understanding of ‘aesthetics’ per se. Context and sense of place, in terms of a setting, are deemed significant either by enhancing or denying good aesthetic appreciation and require an understanding of ‘phenomonological viewpoints’.

In the practice of architecture, the role of the patron is of significant influence, whether the architect’s brief is confined purely to the provision of shelter or whether the architecural product is being exploited to express prestige, power or promotion. As in art, historically religion and secular practices have created the cultural environments that have predetermined the architect’s design inclination. From Greek and Roman times to Modernism, architecture has mirrored the cultural milieu of the era which in turn directs the focus on resources, such as technology and materials that have influenced structural methods and determined the resultant aesthetic.

Down the main stem of architectural history, a persistent thread is evident, that of Nature and natural forms that have inspired both the designer and the craftsman from Greek architecture to the architecture of contemporary times. How Nature has inspired the architect’s work is germane to understanding the intrinsic evidence whether in literal or in abstract terms. Recognised biomimetic studies help to penetrate the often subliminal qualities from which Nature has influenced aesthetic expression in building structures, as well as the facility provided by computer-aided design and the inspiration of non-linear fractal geometrical forms. Thematic iteration in design—similar to thematic transformation in music—is consistently evident in architecture from the Gothic cathedrals to the forefront of the evolution in architectural design today. In considering the above determinants, the picture would not be complete were it not for modern design idioms being assisted by the application of the computer-aided design tools.

Prophetically, in commenting on the new directions in relation to architectural history, Siegfried Giedion (1967) wrote about the third new space conception. He defined the first space conception as the interplay of volumes that can be found in the architecture of Egypt and Greece where the interior is disregarded. The second space conception he observed was started in the midst of the Roman period when the interior and the vaulting system were the highest aim. Giedion elaborated that the third—the new space conception—contains many elements including those from the first and second space conceptions, where the single viewpoint of perspective is abolished thus permitting the “interpenetration of inner and outer space, as well as the interpenetration of different levels. This resulted in an unknown transparency in architecture, which is interpreted by Giedion as Simultaneity” (Wesdijk 2012). Giedion’s understanding of evolution in architectural design concurs with much of how the three space conception is characterised in contemporary architecture with the aid of computer design technology that has significantly advanced into a paradigmatic shift. Defined as ‘parametricism’, a global architectural style has evolved empowered by computational and design technologies such as multi-agent computational systems, generative algorithms and robotic fabrication.

> Architects of the twenty-first century will shape, arrange, and connect spaces (both real and virtual) to satisfy human needs. They will still care about the qualities of visual and ambient environment. They will still seek commodity, firmness, and delight. But commodity will be as much a matter of software functions and interface design as it is of floor plans and construction materials. Firmness will entail not only the physical integrity of structural systems, but also the logical integrity of computer systems. And delight? Delight will have unimagined new dimensions.
> Mitchell (1995)
>
> As the world today struggles with the consequences brought about by climate change, political conflict, population growth and other means by which people are displaced, and urban settings are built and re-built, it becomes increasingly important to consider how we, as designers, architects, and planners might infuse places with richer meaning, poetics, atmosphere and strive for the spiritual essence of space. By establishing a definition derived from characteristics of peak experience, we can begin to translate the ethereal notion of spirituality to a form that has demonstrable implications for the built environment.
> Birch and Sinclair (2013)

In Chap. 2 that follows, the elemental determinants are explored holistically and in more depth for their relevance in architectural design and more specifically to what extent their collective attributes contribute to the 'structural aesthetic'.

References

Balkin, J. M. (2002). *Cultural software – A theory of ideology*. New Haven, CT: Yale University Press.

Birch, R., & Sinclair, B. R. (2013). *Spirituality in place: Building connections between architecture, design, and spiritual experience* (pp. 80, 81, 87). Alberta: University of Calgary.

Charleson, A. (2005). *Structure as architecture: A source book for architects and structural engineers* (p. 22). London: Routledge.

Chiorino, C. (2016). Docomomo Italy member and advisor for the preservation of the architectural heritage for the Pier Luigi Nervi Project Association. Chair Marco Nervi.

Danielsson, C. B. (2011). *Experiencing architecture – Exploring the soul of the eye* (pp. 95–101). Stockholm: The Royal Institute of Technology (KTH).

Dawkins, R. (2006). *The selfish gene* (3rd Rev. ed.). Oxford: Oxford University Press.

Distin, K. (2005). *The selfish meme: A critical assessment* (pp. 2–5). Cambridge: Cambridge University Press.

Fletcher, B. (1948). Gothic architecture 2: Architectural character. In *A history of architecture on the comparative method* (14th ed., p. 346). London: B.T. Batsford.

Gartman, D. (2009). *From autos to architecture* (p. 373). New York: Princeton Architectural Press.
Giedion, S. (1967). *Space, time and architecture, the growth of a new tradition* (p. 5). Cambridge, MA: Harvard University Press.
Gleick, J. (1997) *Chaos: Making a new science* (pp. 100–103, 117). London: Vintage.
Khan, F. R. (1980). Structural aesthetics in architecture and its social and technological relevance. In *Article in Zeitschrift: IABSE congress report AIPC=IVBH, Kongressbericht, Nov.* (p. 136).
Iversen, M. (1993). *Alois Riegl: Art history and theory* (p. 13). Cambridge: The MIT Press.
Jones, P. (2006). *Ove Arup: Master builder of the 20th century* (pp. 195–196). New Haven, CT: Yale University Press.
Loos, A. (1971). Ornament and crime. In M. Bullock (Trans.), *Ulrich Conrad's programs and manifestoes on 20th-century architecture* (pp. 19–24). Cambridge, MA: MIT Press.
Lorenzi, M. G., & Mauro, F. (2010). Art and mathematics in Antoni Gaudi's architecture: "La Sagrada Familia". *Journal of Applied Mathematics (Aplimat), 3*(1), 126.
Mitchell, W. (1995). *City of bits: Space, place, and the infobahn* (p. 105). Cambridge, MA: MIT Press.
Nasar, J. L. (1994). Urban design aesthetics: The evaluative qualities of building exteriors. In *Environment and behaviour* (pp. 377–401). Columbus: Ohio State University.
Nervi, P. L. (1955). Concrete and structural form. *Architect and Building News, 202(27)*, 523–529.
Richmond, S. (1994). Gombrich's critique of aesthetics. In *Aesthetic criteria and the philosophies of science of Popper & Polyani* (11th ed.). Amsterdam: Rodopi B.V.
Rogers, E. N. (1957). *The works of Pier Luigi Nervi*. vi, ix, x, xi. (Preface by Pier Luigi Nervi; Introduction, Rogers E N; Explanatory notes to illustrations, Joedicke, J; Translation, Priefert, E). London: Architectural Press.
Salingaros, N. A. (1999). Architecture, patterns and mathematics. *Nexus Network Journal, 1(2)*, 75–85.
Sparshott, F. (1994). The aesthetics of architecture and the politics of space. In M. H. Mitias (Ed.), *Philosophy of architecture* (p. 13). Amsterdam: Rodopi.
Statham, H. H. (1991). Architecture and dome. In *Encyclopædia Britannica* (Vols. 2 & 4).
Wesdijk, M. (2012). Paper: Modernism and meaning. In *Siegfried Giedion's interpretation of modernism explained by his 'new space conception'* (Subchapters 2.1 & 2.2, pp. 1–3). Eindhoven University of Technology.

Online

Structural Systems. (2016). *Bird's nest*. Accessed November, 2016, from https://beijingbirdsnest.wordpress.com/structural-systems

Chapter 2
Elemental Determinants

Abstract In order to provide substantial argument in the evaluation of the 'structural aesthetic', this chapter explores the most significant determining elements. The true measure of aesthetic experience requires convincing formulation and is explored in the research of recognised commentators on aesthetic appreciation. Apart from cognitive and subjectivity aspects that underly the appreciation of architecture as an art form, the role of more pragmatic factors is evaluated, such as symbolism, the contextual framework, or the setting, together with the attributes of symmetrical and asymmetrical form in which architecture is conceived. History has shown that the role of the patron is paramount in the formulation of the architect's brief, thereby challenging ethical standards. Equal importance is given to the imperatives of religious and secular practices around the globe that to a greater or lesser extent have a bearing on the shape and scale on aspirations that have guided the architects' briefs. As with all enterprises, architecture is influenced by the availability of the materials, methods and resources available economically at the time, thereby becoming important constraints. Nature as mentor or model for form for architectural work is given recognition and expression through evolved forms of geometry including 'fractal' geometry that embodies thematic transformation. By way of the application of natural forms in design, the development of digitally-aided parametric design has pioneered a new approach in contemporary architecture and is recognised as the successor to post-modern and modern architecture. The future of paradigm shifts that are unavoidable include the use of artificial intelligence, of Smart materials, the rise of social media and the interruptive influence of the avant-garde into the established cultural milieu. Collectively these imply new trajectories that are bound to be expressed in architectural design in the future.

2.1 Aesthetic Appreciation and Symbolism

The selection in this book of the illustrious few identified as 'masters of the structural aesthetic' would not be defensible unless at the outset the aesthetic criteria underlying the choice are defined. This suggests the need to venture into philosophical space where art commentators and critics have attempted to

D. Thomas, *Masters of the Structural Aesthetic*, DOI 10.1007/978-981-10-5445-7_2

understand and assign human response to excellence whether in art, architecture or design generally.

Subconscious and subjective responses that constitute aesthetic appreciation have been the focus of a plethora of research by various observers from philosophers and psychologists, to art and architecture historians, to wider fields even of the neuro- and cognitive sciences. What constitutes 'beauty' and 'excellence' and what provides "pleasure in goodness in the physical and social worlds" (O'Byrne 2009) remains elusive—an abstraction as to what visual perception is about. Until the more recent advent of mathematical modelling, empirical approaches to perception yielded merely rudimentary understanding of what perception is (Ciftcioglu and Bitterman 2013). For the purposes of this study it is useful to consider what constitutes aesthetic appreciation of the arts on the one hand, and by extension what constitutes 'the structural aesthetic' in the field of architecture. What features evoke favourable evaluative responses? Design reviewers base their judgements on how physical features influence the evaluative quality of the building and its surroundings and some people may view aesthetics as qualitative and idiosyncratic. Researchers continue in their search for general principles (Nasar 1994).

Aesthetic experiences tend to elude psychological explanation and scientific comprehensive theory and could account for the large number of individual styles, innovation and conceptions on offer in the modern arts. According to the model devised by Leder et al. (2004), "aesthetic experiences involve five stages: perception, explicit classification, implicit classification, cognitive mastering and evaluation. The model differentiates between aesthetic emotion and aesthetic judgements as to types of output."

In his paper on *Aesthetic Measure*, Birkhoff (1933a) conceived a mathematical formulation relating to the aesthetic experience. He observed that: "The typical aesthetic experience may be regarded as compounded of three successive phases, namely: preliminary attention, which is necessary for the act of perception, and which increases in proportion to what we shall call the complexity (C) of the object; then the feeling of value or 'aesthetic measure' (M) which rewards this effort; and finally the realisation that the object is characterised by a certain harmony, symmetry, or order (O), more or less concealed, which seems necessary to the aesthetic effort." To embody these variables in a formula, he was led to write $M = O/C$ that would determine the aesthetic measure by the density of order relations in the aesthetic object.

"Structural engineering is primarily based on logic, reason, mathematics and the understanding of the properties of building materials, and significantly on the forces of Nature. There is very little room for an engineer to indulge in romantic aesthetic forms. But reason in Nature has its own inherent aesthetics. Each building material when used in an efficient, simple and sensitive way, leads to structural solutions which have its visual strength and presence" (Khan 1980a). In recognising the architect-engineer dichotomy—beauty rather than inherent beauty—"aesthetics and technology need to be re-united again." Only by working closely can the structure be expressed honestly without any "conscious effort to mould it into 'art', to contribute to the overall "architectural strength and beauty at the same time." Khan's conclusion is that "Structure is based on reason which has its own inherent aesthetics."

> Thinking of cognition as perception, it is both the fruit of intuition about the whole, where one becomes aware of the general organisation of shapes, colours, place and function in relation to each other, and the intellectual analysis of the elements, operating through the listing of each and every one, and its particular properties.
> Agostinho (2005)

Whereas understanding the human response is germane to the adjudication of architecture of worth, and by association the structural aesthetic, any attempt to find a singular formula for good design is problematic. A more reliable method of discovery of what constitutes aesthetic excellence is the cognitive appreciation of the visual harmony, the social and cultural influences, the technological and intellectual resources at the time of conception, the mathematical application informing the design, the association with natural forms, and not least the refined execution of the final product.

Subjectivity as a human trait is a process of evaluation that is common to all, yet as an measure of excellence it tends to be more from an emotional standpoint and differs from person to person. Being a reflex to individual visual experience, subjectivity escapes ready definition.

For the average person the word 'beauty' covers a range of attributes in art, sculpture, architecture, and even of a natural unsullied landscape. Although adequate as a widely held view of, for example, trends in fashion or the beguiling lines of the latest model of motor car, 'beauty' is rarely adequate where the harmonious integration of physical elements needs to be defined in terms of 'wholeness' or 'unity'. In other words the descriptor 'beauty' does not give adequate recognition to the holistic integration of structural form, function and the face a building presents to the immediate context and the world in general. To describe a fine building as 'beautiful' would undervalue its intrinsic worth. In this instance aesthetic excellence is better served through the understanding of subliminal influences, including the role of mathematics or Nature on the design of a building.

In the evaluation of what constitutes architectural, and by association structural excellence, where the observer returns to fundamental values the time-honoured words of Vitruvius in his book *The Ten Books of Architecture* (probably written between 30 and 15 BC) of 'firmness, commodity and delight' come to mind. John Ruskin's (1849) *The Seven Lamps of Architecture* condenses excellence in architecture into 'sacrifice, truth, power, beauty, life, memory and obedience'.

Significant is the continual bonding between the descriptors 'truth and beauty' in much of the research on the essence of good aesthetics. As it is in the common lexicon, the word 'beauty' appears in parenthesis or in direct references in the text that follows.

Another descriptor that is often applied to the visual arts is that of 'taste', which is defined in the dictionary as the 'aesthetic' discernment in art or literature or conduct. In common usage 'taste' applies regularly to a flavour, a sensation caused in the tongue, or a small dose of a substance to have the experience of it. Applied to aesthetics, 'taste' is therefore too wide-ranging and broad in meaning to apply to the nuanced experience of art or architectural form. Benedetto Croce (2013) disagrees, observing that: "The criterion of taste is absolute, with the intuitive absoluteness of the imagination." However by way of qualification and by stating: "Of course there is as a matter of fact a great deal of variability in critical verdicts...but this is largely due to variances in the 'psychological conditions' and the physical circumstance of spectators", Croce's theory about taste for the purposes of this narrative is not sufficiently definitive in formulating the criteria by which aesthetic excellence should be judged. Croce's concept of 'finality, quality and beauty' has met criticism, but its general tenet that art is "perfectly defined when simply defined as intuition, has proved to be much more illuminating than any previous theory."

In fact, to reflect critically on a work of architecture, the term 'honest' is a more preferred evaluation than the term 'beauty' which, as has been stated, is more evocative than 'taste or style'.

Where Vitruvius (1 BCE), Read (1984), Bakhtin (1990), Pipes (2003) and Croce (2013) set about defining the aesthetic order, Ching (1979) approached aesthetic appreciation through a different lens. He examined every principle of architecture, juxtaposing images that span centuries and cross-cultural boundaries to create a design vocabulary that is "both elemental and timeless and included topics such as point, line, plane, volume, proportion, scale, circulation, and the interdependence of form and space."

Form Finding is the search for optimal shape. In architectural design: "form finding can be seen as shape finding with the optimum being rather subjective to the opinion of the architect. In structural design, the shape itself is not optimised, but the optimum is found in an optimal structural behaviour. For finding the optimal structural shape, different techniques can be used (analogical, digital, algorithmical).
Khabazi (2011)

Vitruvius wrote about "the mathematical basis of architectural symmetry and the importance of geometry for building in general" confirming the later mediaeval belief that there was a "rational, mathematical order in the visible universe." He claimed that "the same mathematical order was to be found in the human body, which could thus be taken as a model for architecture", and "So too in architecture, the mathematical relations among the parts were essential to their design."

Historically architecture and mathematics are two indistinguishable disciplines. "In the ancient world, mathematicians were architects, whose constructions—the pyramids, ziggurats, temples, stadia, and irrigation projects—we marvel at today.

In Classical Greece and ancient Rome, architects were required to also be mathematicians" (Salingaros 1999a), and master masons were enabled through their grasp of geometry to construct their cathedral architecture according to mathematical principles. Even though not written down extensively, this tradition was also the cornerstone of architecture and patterning in the Islamic world.

Subliminal mathematical order as well as the proportions of the human body have a significant place in the evaluation of aesthetic worth that inclines humans to respond more readily to pleasurable sensations that are presented to the senses—thus, a sentient response to shape, mass and surface. Researchers even in neuroaesthetics have addressed the essence of art, believing that neuroscience may be the path to understanding the evaluation in the arts. The enjoyment of aesthetics by individuals has been investigated by using brain imaging experiments. Delegating the role of neurons to the pleasing visual stimuli that are actively derived from horizontal and vertical lines, the aim is to understand the neurological mechanisms involved (Ramachandran and Hirstein 1999).

> Researchers, Ramachandran and Hirstein, developed a highly speculative theory of human artistic experience and the neural mechanisms that mediate it called Visual Metaphors. "Described as a mental tunnel between two concepts that appear grossly dissimilar on the surface, but instead share a deeper connection. Similar to the effects of perceptual problem solving, grasping an analogy is rewarding. It enables the viewer to highlight crucial aspects that the two objects share. Although it is uncertain whether the reason for this mechanism is for effective communication or purely cognitive, the discovery of similarities between superficially dissimilar events leads to activation of the limbic system to create a rewarding process."
> Ramachandran and Hirstein (1999)

The challenge for architects is to coalesce spirituality and place by moving toward a holistic relationship between architecture, people, and the environment. Confronted with images of a particular level of aesthetics, the specific brain areas that are activated can be identified. It is argued that the sense of 'beauty' and aesthetic judgement presupposes a change in the activation of the brain's reward system. Birch and Sinclair (2013a) observe that when considering the aesthetic worth of a building architects and designers tend to talk about "meaning, beauty, poetics, connection, atmosphere and ethereal aspects, and while subjective, these facets of design are no less important than more quantifiable elements such as form, space, and order", but more importantly the imperative should be "a seamless interface between the ordered, quantitative aspects of design and the qualitative, less tangible aspects [that] prove critical to building successful places."

Kant in *Aesthetic Theory: Subjectivity vs. Universal Validity* (Atalay 2007) observes that: "Similarly, the processes of our cognitive faculties that accompany an aesthetic judgment are presupposed in the same way in all humans and these processes must be based on the same subjective principles. Therefore, the aesthetic feeling, whose root is the free harmony of our cognitive faculties on a non-cognitive level, must be universally communicable too. Every person, as a consequence of cognitive faculties, possesses the conditions of free play of the imagination and understanding abilities. In this way, something like common sense is presupposed."

Read (1984) observed that "Any general theory of art must begin with this supposition: that man responds to the shape and surface and mass of things present to his senses" and that "certain arrangements in the proportion of the shape and surface and mass of things result in a pleasurable sensation, whilst the lack of such arrangement leads to indifference or even to positive discomfort and revulsion", even "the sense of ugliness". He concludes that "just as people who are colour-blind are comparatively rare, so there is every reason to believe that people wholly unaware of the other visible properties of objects are equally rare."

On cognition there is an association with being able to build abstractions, and that is the basis for perception and the start of cognition of the structure of an image. In essence the appreciation of the aesthetic or excellence arises from perceived "virtues and character strength, leading to transcendence, one of the six universal virtues. Transcendence and its strengths revolve around allowing individuals to go beyond connecting to others and to connect to the larger universe, which is believed to facilitate a more meaningful life" (O'Byrne 2009). The uplifting experience arising from confronting physical 'beauty', a skill or talent and moral goodness arouses emotional responses of awe and wonder that facilitate a more meaningful life (Peterson and Seligman 2011).

Weber (2008) in discussing the 'myth of meaningful forms' concludes that: "Equally popular is the tendency to define architecture in semantic terms. Like language, it is often asserted that buildings can denote, represent or even communicate earthly matters or metaphysical affairs. . .these are not the dependably specific semantic denotions that are required to make up a language. Thus we cannot speak of architectural communication using these terms."

An observation by Mitias (1977) of the mythical meaning in architecture focusses on whether the "mind does not merely undergo a certain kind of 'feeling', it also knows what sort of feeling it will (should) undergo" and he questions: "Aesthetic perception is not, and cannot be, arbitrary, it is an occasion in which a rich feeling is invested with insight."

From the literature that focuses on what constitutes aesthetic excellence, commentators, have assigned further abstractions. Based on cognition Bakhtin (1990) proclaims that "art celebrates, adorns and recollects. . .it creates the concrete intuitive unity. . .places man in Nature. . .it humanises Nature and naturalises man".

Although intuitive in origin, form can be measured in intellectual terms for balance, rhythm, harmony and association with cultural influences. Responses vary from individual to individual, yet where concurrence occurs, it is likely that it is initially an emotional process, "accompanied by all the involuntary reflexes which a psychologist would associate with an emotion" such as 'empathy'—the attribution to an object, such as a work of architecture, that excites one's own emotional or intellectual feelings about it (Read 1984).

There are other dimensions that have a bearing on visual completeness or integrity, such as the symmetry or asymmetry of form, whether in the second or third dimension. Across fields of science, pattern recognition and knowledge transfer is of relevance to finding geometric arrangement and the sensorial perception of their qualities.

On the question of symmetry, Pipes (2003) observes that it is the attribute that "conveys a sense of stability, permanence, and dignity, as typified in Classical and Neo-classical designs. It could be argued that symmetrical buildings. . .are easier to make and build." In *Analytical enquiry into the Principles of Taste,* (Knight 1972) puts forward the notion that the "beauties of symmetry are only perceived when the building is shown 'from the point of sight at right angles with one of the fronts'; hence for this reason he now recommends asymmetry as being the only way in which buildings will correspond with the natural mountains, lawn and woods." Computer-aided design makes more complex asymmetrical forms possible and is termed by some commentators as 'informal balance' which relies on the creative senses to ensure that the composition is successfully balanced.

Visual harmony by nature appeals to an inner preference for satisfying proportions. The roots of proportion have been studied scientifically since the time of Vitruvius, from which the Golden Section emerges as an ideal ratio defining shape. "A German writer, Zeising, tried to prove that the Golden Section is the key to all morphology, both in Nature and in art; and Gustav Theodor Fechner, the founder of experimental aesthetics, whose principal works were published in the nineteen seventies, made it one of the foremost objects of his research. Since then, practically every work on aesthetics includes some consideration of the problem" (Read 1984). However, rather than accepting that the Golden Section prevails everywhere in works of art, it would be prudent to assume that those in the field of the arts are drawn to it inevitably by an instinctive sense of form, and historically, a wide application of the proportion is found in architecture.

> A mathematical approximation of the celebrated Golden Section is a rectangle of approximately 5:8 proportion. It can be defined as a line that divides a rectangle such that the lesser portion 'a' is to the greater 'b' as the greater 'b' is to the whole 'a' + 'b'. This can be expressed algebraically by the equation of two ratios: a/b = b/a+b. Doczi describes the Golden Section as a "uniquely reciprocal relationship between two unequal parts of the whole, in which the small part stands in the same proportion to the large part as the large part stands to the whole. . .The complete reciprocity of this proportion strikes us as particularly harmonious and pleasing, a fact that has been proven by many scientific experiments.
> Doczi (1985)

Using proportion optimally in design, two objectives should be satisfied, namely, to create pleasing aesthetics through harmonious proportion and to create a sense of order amongst the elements in a visual construction. Euclid held that "a ratio refers to a quantitative composition of two similar things, while proportion refers to the equality of ratios" (Ching 1979).

Where the timeless principles of architecture are expressed through a syntax of shape, line, volume, proportion, scale, form and space that have become the design vocabulary that bridges cultural boundaries, the symbolic association of forms from time immemorial is based on the "simplified abstractions of complex realities, where the power of the visible is signified invisible" (George 2005). Expanding on this deduction, George concludes that symbols "provide a simpler yet powerful means of signifying what one wants to communicate; even spatially, to the watching world. Therefore, understanding symbols and symbolism that is inseparable of any culture holds the key to understand and interpret these. It is only when such interpretations are facilitated that the designer could create meaningful architecture that is of any contemporary relevance. Therefore, training the students to perceive and interpret symbols is quite relevant in architectural education." George associates symbolic attributes with: the point (the beginning of geometrical matter); the line (the extension of the first dimensions displaying length); the triad (the visible dimension, Trinity, displaying fullness); the cube or square (the first solid in three dimensions, displaying length, breadth and thickness); and the circle (without beginning or end, displaying timelessness, perfection and equality).

Another side to the power of aesthetics is that through symbolism, architectural imagery has the power at a denotative level to promote a cause or campaign. As a backdrop image to a campaign to transfer abstract values it can endow power and authority to a campaign. For example, in *Architecture as Political Image*, Cairns 2012a) observes that by using architecture it is possible to promote candidature for high political office against a background image of an august institutional building of state.

Aesthetic appreciation of a building relies on attributes that are pleasing to the observer, however the role of the context or space into which a building becomes an intervention should not be underrated. For a building to enhance, and to be enhanced by its context and mental appreciation, Vali and Nasekhiyan (2014a, b) observe the importance to "recognise place variables and factors that play a role in the creation of place and its sense by considering the phenomenologists' viewpoints and their implied analysis." Structural characteristics play a significant role to make space alive and change it into place and with place sense.

Danielsson (2011) explores the 'poetics of space' written by French philosopher Gaston Bachelard, who presents for consideration "two perspectives in understanding the influence of architectural settings on people: the phenomenological perspective—structures of experience and consciousness—and that of neuroscience." A phenomenological inquiry is deemed necessary in order to understand the architectural experiences since this approach includes our "sentimental resonances by which we receive a work of art" (Bachelard 1994).

Developing this concept further, Danielsson (2011) observes "that all sensory senses form the basis of how architecture is experienced,. . .includes mind, soul and body." The phenomenological aspect is thus significant in the cognitive appreciation of the architectural element as it exists within, and relies on its context. Danielsson avers that there is, however, "a single-minded focus on the visual side of architecture that already suffuses architectural education."

Where aesthetic theories contribute to a better understanding of the qualities, or determinants, that underpin the cognitive aspects of architectural appreciation, studies in cultural preference, cultural proclivity, or natural inclination, allude to a behavioural phenomenon that spreads from person to person within a culture, that became the focus of a field of study called 'memetics'. In the abstractions of perception and cognition, experience and meaning, and human response to the aesthetic attributes of the visual arts and architecture, since the early 1990s researchers and commentators have added this consideration: the cultural dimension, or the collective mindset. Studies in 'memetics' have explored cultural proclivity, or natural inclination, a behavioural phenomenon that spreads from person to person within a culture.

In *Cultural Software: Theory of Ideology*, Balkin (2002a) states: "Individuals embody cultural software and spread it to others through communication and social learning. Ideology is neither a special nor a pathological form of thought but an ordinary product of the evolution of cultural software." This line of philosophical argument, or the collective mindset, has been further developed by Richard Dawkins in *The Selfish Gene* (2006a). In terms of an evolutionary model and transmitted from one mind to another, some researchers consider memes as analogous to genes, self-replicating and responding to environmental influences. The term 'meme' was coined by Dawkins in 1976 which derives from the Greek word mimema (something imitated). The relevance of this study is that it resonates with the cultural trends in communities, even nations, in many fields such as in art, music, philosophy, writing, fashion and lifestyle. Dawkins likens the process by which memes survive and change through the evolution of culture to the natural selection of genes in biological evolution. Balkin's theory (2002a) of 'cultural software' maintains that memes form narratives, networks of cultural associations and a variety of different mental structures. It is an interesting philosophy to begin understanding more about how the dynamics of cultural are expressed in the living environment, and by association in architectural style and even the ordering of public space. The collective mindset is evident in all indigenous cultures down the Ages: the timeless vocabularies of architectural style such as that of Classicism or Gothic, or the vernacular styles of Russia, the Middle-East, and China to name only a few.

Theories of what constitutes 'the aesthetic' have occupied a large space in academic discourse in attempts to define the soul of aesthetics, which remains elusive and accounts for the multi-dimensional interpretations presented. Those theories that illuminate have become the touchstone for the selection of the ten 'masters of the structural aesthetic' represented in this book.

2.2 Context and Sense of Place

For a proper evaluation of true architectural worth, the context or the space in which as building resides is as important as the architecture itself. Buildings are meant to be context dependent, the term 'context' being understood more widely in terms of both physical, i.e., urban space, roads, neighbouring buildings, land contour, the surrounding landscape both natural and man-made, and the non-physical elements that would include cultural, political, green environmental and economic constraints.

"The completeness/unity that the context prompts us to use in our work space in an effective manner, allows us to clearly identify solutions to any problems that we face and their successful resolution; and is useful for the evaluation of the elements that belong to it. The importance of context emerges when its ability to express and bring to light an item's contents, its inclusion within a combined unity, its definition and its formation and attainment of a meaning is the matter of discussion. Thus, the item becomes an element of the entirety/whole and becomes an integral part of the context" (Çizgen 2012).

Architecture is inevitably intertwined with contextual thinking, where on the one hand the architect allows the design to be informed by the context, and otherwise where the design objective is to transform the context or surroundings. Shape assessment and scale determine the dominance or subordination of an architectural intervention, the former by which means a banal urban environment could benefit—the shape diverging optimally from the prevailing urban fabric—and the scale expressing the stature of the building in relation to its immediate or wider context. If aesthetic character is contextual, this would be largely due to an amalgam of cultural and historical preference, and be reliant on available local materials and construction techniques.

Some commentators are critical of "isolated masterpieces of any sort thus detached from their settings, the poignancy of the contrast between them and what adjoins is too liable to produce a depression that the grandest sight cannot utterly dispel" (Johnson 1994).

A broad discipline of study, phenomenology, is a method of inquiry in philosophy, developed largely by the German philosophers Edmund Husserl and Martin Heidegger, which is based on the premise that reality consists of objects and events (phenomena) as they are perceived or understood in the human consciousness. Architecture and its role in 'sense of place', from a phenomenological viewpoint, links with the concept of a place through the understanding of symbols and daily activities (Vali and Nasekhiyan 2014c). To fulfill the maxim: 'the whole is the sum of the parts', requires the symbiotic integration of both the setting and the intervening elements that need to become inseparable parts of the entirety. This suggests that 'context' should give direction to design by acknowledging that it plays an important role in the conception of good architecture. In the ten case studies in this book, the role of context, the setting, is found to be a compelling determinant and integral with their aesthetic accomplishment.

2.3 Patronage and Ethics

2.3.1 *Historical Eras*

Down the Ages architecture, with its attendant philosophical base, design expression and constructional method, will inevitably be a reflection of the milieu in which it is conceived and executed. Furthermore, the instigating agent from ancient times to present day has relied on the unfolding of socio-political and religious forces characterised by: the primary needs of primitive man; the aspirations of pagan gods; the patronage of the church; the present day public and private patronage; or the cultural amenity needs of the community. The role of the private patron or client in modern times is typically becoming more engaged with stylistic expression and purpose that chooses architecture for the promotion of social image. "The directness of how these concepts and strategies are applied in the political realm, and the way this often appropriates architecture into the political communicative tapestry. . .the role architecture currently plays in political campaign imagery [is] a role which clearly draws on a cultural interpretation of buildings that echoes across continents" Cairns (2012b). This reflects a disposition based on fundamental values specific to a culture, a movement, a place in time.

On the issue of moral or ethical questions relating to structural honesty, Milne (1997a) observes that "There have been and there still are many proponents of the idea that for every architectural/structural problem there is a 'correct', a 'right' and an 'honest' solution. Structural honesty has been linked with the functionalism of the modern movement in architecture with all its social and moral overtones." However, all things being considered, the real test is whether the structural integrity of the iconic buildings being presented in this book has provided them with singular status as 'honest' responses to a quality of timelessness and the extent to which they contribute to the main stem of architectural history.

History has shown that whether state, church, corporate or private, the role of the patron is paramount. The creativity of the designer, builder or craftsman can be overpowered by the pressures of expediency, prestige, politics and potency that exist in the cultural environment of the times. There is another side to the coin, that is, when to decline a request to design a structure which, in the opinion of the designer it is inappropriate, is unnecessarily expensive, or is too difficult to build—in other words it is an apparent wilful demand of the client regardless of ethical considerations. There is also the dilemma that the designer may wish to achieve certain self-indulgent, self-promoting or other ends in fulfilling the client's brief (Milne 1997a). This is as relevant today as it was formerly when the mason, carpenter, engineer or architect was the master builder/designer.

Where Egyptian temples were a royal prerogative, Greek temples were the people's patrimony. Of this symbolic connectedness there are many historic clues: in the Greek period the government of the people was concerned with *res publicae* (the people), edified in the perfection of the Parthenon and the agora, the public place, whereas in the case of ancient Rome, the importance of the State was reflected in the public buildings. Dameron (1991a) observes that it is "nearly impossible, however, to determine a specific origin of the tradition. In fact, "Cicero thought that the origins of Roman clientela were so ancient that it must have been brought to Rome by Romulus himself. By nature it developed very gradually over long periods of time as different families and individuals rose and diminished in prominence in their respective corners of the continent."

In Gothic times "Patronage was practised as a social institution throughout early modern Europe, probably peaking in importance between the fourteenth and fifteenth centuries" (Dameron 1991a), when the outburst of architecture during the Medieval period resulted from the patronage of the wealth and power of the clergy. Simultaneously, "the Monastic Orders made the Church the one great advancement of the Middle Ages, manifest in church building of the thirteenth century when all classes of craftsmen worked continuously. A new method of construction was evolved in which small stones were held together in equilibrium, and the pointed arch became an outstanding feature of the style" (Fletcher 1948a).

The question of the participation of the common people in the building of Gothic churches has received the attention of historians who question whether their pious enthusiasm induced them to assist physically in their construction. "Modern research on the other hand, has viewed with scepticism all accounts of this nature, maintaining that the occasionally hysterical pitch of the faith of the masses was encouraged and precipitated by irresponsible priests and monks interested in channelling this dumb devotion into monetary donations and free services" (Frisch 1987). Nonetheless, according to Coulton (1928), "it would be historically wrong to dismiss as meaningless all church annals speaking of the concern and participation of the local populace in the important events of their cathedrals and churches."

In Florence in the fifteenth century craftmanship and became the backbone of industry. The city boasted 54 workshops for marble and stone; it employed 54 master gold and silversmiths, and at least 30 master painters. The wool and silk industries enjoyed the highest reputation for quality, attracting the patronage of discerning merchants and financiers. One of the defining characteristics of Florentine society was a deeply-rooted culture of patronage, where ownership of land determined one's importance, but despite the prevailing political rivalry on the part of church officials, Italian patronage was largely due to the generosity inherent in Catholicism.

Patronage become instutionalised in Florentine life. As Biagioli describes it, patronage was not an 'option.'. It was the key to social status, and, in Florence, there was an absolute social hierarchy. A career and social mobility were impossible to separate from being involved in a network of patronage relationships. Even the working poor found themselves a part of this complicated web in their labour under Florence's multitude of patron-driven guilds Dameron (1991a). It was therefore not surprising that in Florence, early patronage associated with the church was the result of powerful episcopal and political influence in central Italy.

Private patronage marked a change from before the time of Antoni Gaudi to the present day. Gaudi formed a close association with textile manufacturer, Eusebi Güell, who eventually became his only real supporter and to whom many of his masterpieces can be attributed. Both men were profoundly Catalonian, and lived in a period, "La Renaixença," in which Catalonian language and culture were being promoted. Both were simple people who, and although they were famous and well known in the city, did not show off their wealth (Güell) or genius (Gaudí).

The construction of "La Colonia Güell" for Gaudi's main patron Eusebi Güell ceased altogether when, in 1918 his patron died. Gaudi became introspective and concentrated on his masterpiece, the La Grada Familias. However, the generosity of other wealthy patrons gave his artistic freedom a secure foundation by their trust.

2.3.2 Modern Era

It remains likely that architecture will continue to be exploited as a vehicle by individuals who, through their patronage, commission image-building architecture as a demonstration of their social, or political or financial stature. Writing about architecture in *The Edifice Complex*, Sudjic (2006) observes that there may be no fixed political meaning to a given architectural language, but that does not mean that architecture lacks the potential to assume a political aspect. Few successful architects can avoid producing buildings with a political dimension at some point in their career, whether they want to or not. And almost all political leaders find themselves using architects for political purposes.

Community architecture that serves the social and cultural needs of resident populations is a growth area for institutional patronage in democratic societies and accordingly generally freed from the promotion of political or self image. The rise of world sporting bodies as promoters of mega-sports events has presented economic and social upliftment to hosting cities, and has contributed globally to a wealth of emblematic architecture.

In contemporary terms there is a further dimension that informs a movement to return a city's pride. For example, evident in a former Soviet state, Azerbaijan in the Caucasus, where with the plenitude arising from new wealth from one of the world's largest oil and gas projects, the city has experienced a significant influx from rural areas. In response to this real scenario, politicians have distanced themselves from their Soviet past of political rule in realising the architectural ambitions of the capital city of Baku today. While those trained in the Soviet tradition in the service of state power embrace futuristic buildings a new form of social exclusion could arguably apply at a cost. The World Bank and others have observed the "absence of economic diversification in a country dependent on an oil supply is now understood to be in decline. In this context, what we observe is decidedly familiar: a government's invocation of a certain material strategy to exorcise the past and to compete on a world stage. Indeed, this celebration of glass and steel with the latest architectural intervention, the Heydar Cultural Centre, conjures up the image of the 'technological sublime'—a world where the skyline carries the citizens' imagination above and ultimately away from the known life of the sidewalk" (Marx 1964; Gell 1992).

For the future, it is an open question whether the linking of political idealism should remain a main driver behind architectural, expression or as a true reflection of the meaning of all art. A political agenda should not, cannot be considered a stand alone criterion in reflecting on the honesty and integrity of what informs 'the structural aesthetic'

2.4 Religion and Secular Practices

Down the annals of architectural history the need to experience a deeply meaningful and heightened sense of spirituality has been the driving force in the design of places of worship. Spirituality holds different meanings for different people and cultures, in the spheres of both religious and secular practices and in places that shape, connect to, and engage with a sense of self-actualisation and personal meaning, "The word 'spirituality' holds a different meaning for different people and both religious and secular directions" (Birch and Sinclair 2013b).

Maslow's view on self-actualization and self-transcendence compels the argument that the 'spiritual in design' should be pursued as it has the potential to "inform, inspire city planning, architectural design, place-making, and space-making" (Sinclair 2011).

2.4.1 Greek and Roman

The Greek religion was in the main a worship of natural phenomena, of which the gods were personifications, and each town or district had its own divinities, ceremonies and traditions (Fletcher 1948b). Worship of primitive forms, worship of ancestors and deified heroes were not unknown, and was carried out in rites in which both men and women officiated.

The practice of Roman state religion was manifest in ritual, rather than individual belief, collective rather than personal. Regular animal sacrifices, festivals and offerings to favour the gods were expected to bring favour to the state, household or individual. Generally sacrifices took place on an altar in front of the temple.

In both Greek and Roman times, group ritual worship did not take place indoors as in modern churches, mosques or synagogues, instead religious structures were storehouses for a statue of a god, and other ritual accoutrement connected with a religious place, or practice.

2.4.2 *Gothic*

The rise of Christianity was a striking phenomenon which spread throughout the civilised world, and down the Ages has inspired some of the greatest architectural monuments. Where the temples of the pagan Greek and Romans were to shelter the statues of the gods, the purpose of the Christian church has been to shelter the worshippers.

As a background to the rise of Christianity, Constantine (313 CE) decreed that Christianity and its worship should be given equal rights with other religions. He himself professed Christianity which became the official religion of the Roman Empire. Christians began building churches which fitted their religious needs, and to the glory of God. Thus liberated, the Christian church continued until the present day to provide an appropriate House of God wherever Christians worship.

Imbued with the spirit of God, the Gothic masons strove to build structures that employed structural methods with the resulting aesthetic surpassing all architecture of the past Ages—the structural parts of the building ceased to be its solid walls, and became a stone skeleton comprising clustered columns, pointed ribbed vaults and flying buttresses. Milne (1997b) observes that: “A kind of religious compulsion must have been present to spur these builders to such heights—literally and metaphorically…It has been suggested that Gothic builders set out to create pretentious self-imposed structural problems. But that is to ignore the underlying theme which subordinates engineering means entirely to aesthetic aims.”

2.4.3 *Renaissance*

The consequence of the spread of printing in Europe and England between the fourteenth and fifteenth centuries was the spread of knowledge that engendered a spirit of enquiry and freedom of thought which prompted a breakaway from ‘Romish’ influence. This was characterised by renewed vigour in intellectual life, the Reformation in religion, to the Renaissance in literature and ecclesiastical as well as domestic architecture. The new departure saw the employment of the Classic Roman ‘Orders’ of architecture in novel combinations to meet the predelictions of the day and which had been dormant for a whole millenium. “The arts were in the hands of skilled craftsmen, goldsmiths and workers in metals who looked on architecture as an art form rather than of construction” (Fletcher 1948b).

It was in the spirit of the Age that Brunelleschi conceived and constructed his masterpiece, the Santa Maria del Fiore, the Duomo, the Cathedral of Florence (1420–1434 AD).

2.4.4 Modern Era

Yielding to a new ethos where industry, artists and craftspeople came together, the 'Deutsche Werkbund' (German artwork union) was formed in 1907. Socio-political and economic forces combined to reform German craft and even up to 1933 the Werkbund had a great influence on modern architecture in Germany. Arising from the various architectural currents of the time, industrial production began to overtake traditionally handcrafted methods which included architecture. These same principles led to new forms and materials instead of historical ornamentation or floral decoration, and eventuated in functional buildings where constructional elements were unadorned.

The new industrial culture embraced the use of glass, epitomised in the work of Walter Gropius and Adolph Meyer's Fagus Factory (1911–1913). Relieved of their structural function, the external walls were curtain walls of glass, giving the building sense of lightness, heightened by the all-glass corners. In 1919, Walter Gropius took over the leadership of the Weimar Art School and united it with the former Art and Craft School with a view to create a new unity between art and craft. In the process the aim was to join all the creative forces and to give socio-political meaning following the catastrophe of the WWI and to build a new society (Tietz 1999)

In the period 1900–1920, modern art and architecture became more relevant, backed by social purpose. Housing developments were carried out with urgency to address the problem of homelessness. Within this milieu, the signature works of Gropius (1883–1969), Van der Rohe (1886–1969), Le Corbusier (1887–1965) denoted the arrival of Modernism, a new architecture, not confined to region or cultural expression, that became a worldwide development. Individual responses were characterised by the works of Nervi (1891–1979), Niemeyer (1907–2012) and Otto (1925–2015), later to include freedom of expression in the work of Utzon (1918–2008) and Hadid (1950–2016).

2.4.5 The Avant-garde

Artistic revolution, the practice of a revolt that transgresses the borders of the art world, challenges the dominant political and economical conditions of human existence, and aims to be at the service of political or social revolution, or at least of political and social change, and can function as an inspiration and model for existing art practices. "Today, the political role of art is mostly seen as being twofold: (1) critique of the dominant political, economic, and art system, and (2) mobilization of the audience toward changing this system through a Utopian promise" (Groys 2013a).

Such secular movements tend to assert innovation in design, technology and architectural form, limited only by available skills, materials and resources.

2.5 Materials, Methods and Resources

From the time of the first structures erected by humans for their purpose, reliance on available material resources was inevitable. The constraining use of a material would determine the architectural form. Building a structure from the unit, such as stone or brick, would differ markedly from building in a homogenous material, such as concrete, or steel where the capacity is mostly employed for large spans or the skeleton of a structure. In structural history, the role of unit construction determined the solid, layered character of the Egyptian temples and pyramids, but continued its evolutionary role with greater refinement in the structures of the Roman and Gothic period, that allowed architecture to rise to heights of structural performance that inspired ingenuity and innovation of method and form.

Structural materials for the Pantheon and the Colosseum have been the subject of continuing study by geologists, archaeologists and engineers seeking to unravel the mystery of ancient Roman concrete and it longevity. Roman concrete, not as strong as the modern product, differed from that in general use today and weaker on the scale of approximately ten times. However, where modern concrete is a mixture of lime-based cement, sand and water with gravel aggregates, the durability of Roman concrete against the elements may be due to a key ingredient, volcanic ash. According to the writings of Vitruvius, the first-century BC architect and engineer, the Roman mix of volcanic ash reacted with the lime paste to create a durable mortar combined with fist-size chunks of bricks or volcanic rocks that were packed firmly between shuttering (Wayman 2011a). Stone was clearly an important material for construction and the Romans were skilled in quarrying the material and using different types of stone for different applications, such as marble for superior-finished construction, whereas lime and sandstone would be used for areas of light usage, such as for walkways. Basaltic lava or granite were employed for areas of greater wear.

With the development of the vault, the arch and the dome, brick or stone units relied on structural compression in order to span over space. Steel, due to its tensile property, emerged only later as a contender for structures where lightness, or even impermanence, was a pre-requisite. Available material affords the structural engineer design opportunities, presupposing the knowledge gained of the limitations of structural materials through experience and observation. Master builders were those who had to acquire that knowledge and the skills necessary for working with available materials, from lessons learnt through disastrous trial and error results (Addington and Schodeck 2005a).

With the advent of the Industrial Revolution the place of materials brought about significant changes in the approach to architectural design. The widespread introduction of steel permitted long-spanning building elements thereby liberating the architect and engineer from the constraints of brick, stone and concrete. Improved glass-making technology gave impetus to the 'international style' that in turn allowed facades to become cladding membranes where the aesthetic of the structure was concealed. Accordingly, where advanced climate-control methods catered for matters of occupational comfort, architectural design was less concerned about the structure forming the bones of the building.

Since Egyptian times and until technological advances of the Industrial Age, site construction relied on moving and placing building elements using man-power as the norm. Thereafter with the development of hydraulics fulfilled the functions of lifting and placing, to facilitate the erection of structures and widen the scope and scale of constructional methods. Modern builders can rely on, for example, a configuration that includes: control of construction activity from a centralised computer; satellite computer controllers; double-acting high pressure hydraulic cylinders; and electronically controlled hydraulic power units, custom-designed for the job in hand.

> The role of materials changed dramatically with the advent of the Industrial Revolution. Rather than depending on an intuitive and empirical understanding of material properties and performance, architects began to be confronted with engineered materials. Indeed, the history of modern architecture can almost be viewed through the lens of the history of construction materials. Materials have progressively emerged as providing the most immediately visible and thus most appropriable manifestation of a building's representation, both interior and exterior. As a result, today's architects often think of materials as part of a design palette from which materials can be chosen and applied as compositional and visual surfaces.
> Addington and Schodeck (2005a)

In the middle of the twentieth century, architect Pier Luigi Nervi's understanding of the behaviour of materials and structure was profound, leading to the recognition not only of his peers, but also the august bodies who reward and applaud excellence in architecture. No better testimony would be necessary to justify Nervi's place as a master of the structural aesthetic. "It can be said that the most characteristic architectural forms of reinforced concrete in these last few years have been at once the cause and the consequence of a widespread orientation towards genuinely constructional architecture, which has very quickly reached every country and every aspect of construction" (Nervi 1955).

Some observers endorse concrete for its permanency, and 'poetic finish' and for becoming the 'new liquid stone of our Age'. Also, "Today the possibilities are endless as one can mould, texture, colour, tool and stain concrete to create any desired architectural effect,"...and "concrete as a material can forge new identities through commissioning its use in public buildings" (Van der Merwe 2013).

Glass fibre-reinforced concrete consists of high-strength, alkali-resistant glass fibre embedded in a concrete matrix. In this form, both fibres and the matrix retain their physical and chemical identities, while offering a synergistic combination of properties that cannot be achieved with either of the components acting alone. In general, fibres are the principal load-carrying members, while the surrounding matrix keeps them in the desired locations and orientation, acting as a load transfer medium between the fibres while protecting them from environmental damage. Incorporated into a matrix either in continuous or discontinuous (chopped) lengths, glass fibre provides superior reinforcement for the matrix.

A most significant leap in the field of material resources is the advances in Smart and Nano materials. Where down the Ages architectural character has relied largely on materials available for structural purpose, for the future the "new Smart and Nano materials have the potential to open a new era in architectural design and construction, enabling architects (and engineers) a higher level of intricacy which will span from the smaller scales of a molecule to the larger concepts of society" (Daveiga and Ferreira 2005a).

Material products from Nano technology are materials with a nanometer scale substructure. (The term 'nano' derives from the Greek word for 'dwarf', meaning one billionth of a metre). Manipulation at an atomic level is enabled making it possible to "build the way Nature does it, atom by atom and molecule by molecule, i.e., self-assembly" (Daveiga and Ferreira 2005a). If deployed these innovative advances in materials could profoundly influence the direction of architectural and structural form in the future, however, this would depend on the combined expertise of the architect and the material scientist embracing nano-technological advancements.

In architectural terms, the knowledge gained over time from the behaviour of materials and construction methods has liberated the imagination of architects and builders and has prompted the design of buildings which not only satisfy utilitarian needs, but also the sentient human response to their aesthetic attributes. With growing confidence, creative imagination has inspired iconic building styles that populate the main stem of architectural history from the Egyptian era to the present day, from the hands-on craftsman to the advancement of science and technological skills in the Digital Age.

In his lecture to a meeting organised by the Institution of Structural Engineers and Joint Committee on Structural Concrete on 15 October 1955 (representing the Cement and Concrete Association, the Prestressed Concrete Development Group and the Reinforced Concrete Association), Nervi outlined the contribution that reinforced concrete has made to the development of present day tendencies in architecture. "After the first unfortunate attempts to adapt the new material to the structural forms of masonry or timber, reinforced concrete, spurred on by technical requirements and by its own unlimited possibilities, turned very quickly to new structural forms, which to the surprise of their inventors themselves, were found to possess an inherent beauty of their own."

2.6 Nature, Biomimicry and Biomimetics

Nature as a 'mentor', as a 'measure' and as a 'model' through authentic or honest replication of natural harmony and order, has the potential to appeal to the sense of being integrated and unified with a place or a building. "Biomimicry is defined as the mimicking of life using imitation biological systems" and "presents a unique channel through which an individual can sense fusion with the world and move toward an egoless, self-transcendent perspective" (Kellert and Wilson 1993). It is the one-ness of working in tune with, rather than exploiting, that brings Nature closer and facilitates the peak experience in the individual.

The 'Biophilia Hypothesis' explores humanity's innate affinity for the natural world. In his landmark book E O Wilson examined how our tendency to focus on life and lifelike processes might be a biologically based need, integral to our development as individuals and as a species. That idea has caught the imagination of diverse thinkers and brings together the views of some of the most creative scientists of our time, each attempting to amplify and refine the concept of biophilia.
Kellert and Wilson (1993)

Biomimetic architecture (bios, meaning life and mimesis meaning to imitate) is a contemporary philosophy of architecture that seeks solutions for sustainability in Nature, not by replicating the natural forms, but by understanding the rules governing those forms. It is a multi-disciplinary approach to sustainable design that follows a set of principles rather than stylistic codes. It is part of a larger movement known as biomimicry, which is the examination of Nature, its models, systems, and processes for the purpose of gaining inspiration in order to solve man-made problems (Root-Bernstein 2001).

From earliest times mankind has at various stages sought inspiration from Nature for built structures. Fundamental to the Egyptian culture was the interpretation of Nature in their buildings, evident, for example, in the temple of El-Karnak, where the design of the columns is not only analogous to that of reeds, but in terms of their close spacing. Creating the essence of the natural world was considered to be a gesture to the gods. The similarities between 'the structure in Nature' and 'man-made structures' are evident in the structures of early man, in the designs that display functionality, optimisation and resource efficiency—the minimised use of material and energy. "Thus, the structures of Nature have the capacity to endure internal and external forces in an optimum way [that] are always a source of inspiration for many progressive architects and engineers" (Arslan and Sorguc 2004).

Later, the Greeks articulated and refined their architecture in their strive for perfection. A veritable 'kit of parts', the Orders of Architecture, namely the stylobate, shaft, capital, architrave, frieze and pediment, each in their own way metaphorically expressed Nature yet simultaneously represented a specific structural purpose.

The Gothic architect intricately extended static laws into delicate, iterative themes expressed throughout the main structure, the pinnacle, the flying buttresses, essential to the stability of the whole where components were not pure decoration, but an extension of a basic theme that undergoes constant transformations and disguises and is made to appear in several contrasting roles. "The branched support tree-like structure can be first observed in the ribs of Gothic style. Today, tree-like structures are three-dimensional supporting systems used increasingly in steel, wood, and concrete buildings" (Arslan and Sorguc 2004)

In the neo-Classical period the true structure of a building was invariably fully or partially enclosed in ornamented cladding conceived independently of the parent structure. Ornament became an anathema to the Modernists who deliberately designed buildings shorn of stylised embellishment, and argued that aesthetically, unfinished concrete was intrinsically desirable.

Nature continues to provide a virtually unending range of functions and materials to inspire adaptation, even for damage repair and self-healing. Purely observations in Nature will gradually be replaced by a systematic approach involving the application of engineering principles to further development of bio-inspired ideas and the generation of specific databases (Frazl 2007a).

"The modern emphasis on honest structural expression is more sincere, with innovative interpretations of Nature in spatial structures, examples being the exemplary Nature-inspired tectonic architecture of Jørn Utzon, together with the works of others, such as those of Gaudi, Candela, Otto, Calatrava and Nervi" (Carter et al. 2013). Carter observes that to give effect to the "direct inspiration of Nature" is the "increasing use of parametric digital design tools that replicate virtually instantaneously evolutionary processes resulting in structures that are not only elegant tectonically and in terms of economy of means, but also aesthetically pleasing, profoundly satisfying experientially and often culturally significant."

Some Modernists have gone further and conceived buildings where the circulation and services were external to the building envelope, such as the work of Rogers' Pompidou Centre, Paris and Lloyds of London. Louis Kahn gave identity to this typology describing the design theme as 'served and serving elements'.

> The concept of biomimicry uses Nature as a "mentor", Nature as a "measure", and Nature as a "model" (Benyus 1998).

Nature displays visual qualities from animate and inanimate forms that are expressed dynamically in structures that are shaped by internal and external forces. "Nature's patterns and sequencing of systems that form part of the physical and biological processes of the natural environment create a form of organic architecture used traditionally by indigenous groups, and some architects, including Antoni

Gaudi. . .to speak to the inherent biological affinities of the human being. Fractal geometry [dealt with in the next subsection] illustrates Nature's intricacies and alludes to its internal processes and thereby are windows of opportunity for designers and other disciplines to learn from the elegant efficiency of the natural environment" (Ahuja 2016).

A fitting example of a biomimetic structure in Nature is the termitary, the towering mound of the *Macrotermes bellicosus* (formerly *nataliensis*) termite, a living ecosystem and model of energy efficiency. In 1991, Zimbabwean architects (Pearce et al.) together with the project engineers (Ove Arup) adopted the design principles of the termitary at the Eastgate building, Harare.

The towering mound built by *Macrotermes bellicosus* is really a giant ventilation chimney through which hot air from the nest can escape. Beneath the tower is a cave, typically about three metres in diameter, housing the nursery galleries, the queen's cell, and the fungus gardens. Below the main cave are cavities, ten metres or more deep, from which the termites obtain water. At the top of the main cave is a hole, which the termites can make bigger or smaller by adding or removing soil particles. This varies the speed of warm moist air passing up and out through the cave and chimneys, and controls the temperature to within one degree.

Studies of termitaries of this species on the Ivory Coast of Africa show that these insects have established an ingenious structural system to serve their accommodation needs. A cross-section through the mound shows the nest proper, which is almost round, with its royal cell in the centre, and its many chambers and passages. There is a larger space, the 'cellar', while the central structure rests on conical supports and is further anchored by lateral struts. Another air space above it reaches a long way upwards like a chimney. On the outside of the mound, ridges and buttresses run from top to bottom and radiate from the upper air space into the ridges, where they divide into small ducts. These come together again to form channels as wide as the first leading into the cellar. Of these structural solutions, many are found on formally constructed architectural forms.

Present day technology allows designers to better visualise and optimise the multiple fractal combinations found in Nature to create more resilient and smarter buildings. Jacob Bronowski, physicist, poet and humanist, observed: "The discoveries of science, the works of art are explorations—more, are explosions, of a hidden likeness" (Root-Bernstein 2001).

Biophilic design also incorporates the use of fractals or Nature's patterns and sequencing of systems that form part of the physical and biological processes of the natural environment. This creates a form of organic architecture used traditionally by indigenous groups, and to some architects, including Antoni Gaudi. . .speak to the inherent biological affinities of the human being. (Ahuja 2016).

Consider the individual tree, and its dynamical system that it owes to unique circumstances, such as soil composition, gravity, climate, predators amongst other

Fig. 2.1 Natural forms: Tree structures; Rock formations; Termitaries, spider webs and other forms of animal constructions have inspired architectural forms

influences, it is Nature's gift to strength and the structural aesthetic (Fig. 2.1). It is commonly deployed from historic to contemporary architecture as inspiration for the design of supporting structures. Arslan and Sorguc (2004) noted that Antonio Gaudi, who practically never journeyed anywhere, drew his inspiration from his ability to observe and reuse the countless details offered by Nature. He invoked the tree as a source of inspiration for his dendriform structural columns that support the intertwined hyperbolic vaults and mimic the branches of a tree, strikingly analogous with the four inclined basalt columns of the Passion Façade of the Sagrada Familia in Barcelona—the Façade provides the sensation of an organic and natural structuring, like the trees in a forest.

As a biologist, the question for Benyus is not whether our technology is natural, but how well adapted it is to life on Earth over the long term. She says that engineers, scientists, architects and designers are often humbled, and then excited, when they discover how Nature already has solutions to their challenges, and how it generally outperforms their traditional solutions, showing them creative alternatives. 'Nature knows what works, what is appropriate, and what lasts here on Earth.'
Benyus (2002)

Arguably in more contemporary examples, a wider evocation of Nature in architecture is acknowledged as a valid paradigm, flowing from a conscious adaptation of the shell, a natural form. Capable of metaphoric interpretation and abstraction, natural forms have inspired some of the most iconic structures, such as

Jörn Utzon's the Sydney Opera House, that opened in 1973, that has been amongst the most widely acclaimed structures of the twentieth century. It continues to capture the public's imagination. The structure is constantly referenced to natural form, and follows Utzon's own personal architectural credo, *The Innermost Being of Architecture* written in 1948, where Utzon states that, "The true innermost being of architecture can be compared with that of Nature's seed, and something of the inevitability of Nature's principle of growth ought to be a fundamental concept in architecture" (Weston 2002).

Initially Utzon had not been able to achieve a satisfactory solution to the realisation of the paraboloid egg-like roof structures. His eventual breakthrough came with the adoption of a spherical geometry, Nature's way of containing the greatest volume with the least surface area.

Capable of covering large spans economically in concrete, there is comfort in the aesthetic association with the egg, the seashell, the scallop, the shell of the turtle. Designers assisted by mathematical and geometrical electronic aids have become equipped to explore shapes and forms that express strong organic links, and attain a striking structural aesthetic.

Utzon himself had admiration for the work of Gaudi, the underlying structure of whose architecture is remarkably elegant and daring in terms of its engineering, based upon a profound understanding of how forces operate in Nature, analogies to natural forms, particularly the bee-hive and the use of weighted string models (Carter et al. 2013).

Other innovative typologies in architecture have counterparts in Nature. Frei Otto's tent-like structures have exploited the natural structural form which relies on tensile strength and exploits the catenary curve for efficiency and economy for spans over large spaces. The structural principle he adopted is analogous to the strong and elastic silk-like web form of the spider that is capable of extraordinary load bearing capacity. Otto's landmark tensile structures covered with a lightweight membrane sheltered the seating at the Munich Olympics of 1972. Designed in collaboration with the architect Günther Behnisch, the Nature-inspired structural form, in essence the traditional tent applied to modern structures, entered architectural history adding significantly to another form of the structural aesthetic.

Of more recent significance, in terms of a strong reference to Nature, is the work of Zaha Hadid where Søberg (2016) observes, "Along with her strong conceptual and historical awareness, Nature's forms and shapes appear as a recurrent source of inspiration for Zaha Hadid's architecture. It includes attention to physical contexts and landscapes, whether resulting in layered structures or powerful moving lines but also exploring possible interfaces between patterns and construction".

2.7 Fractal Geometry

Down the Ages, Euclidean linear geometry appeared to dominate architectural form, where properties such as symmetry, the rectangle, the triangle and the circle determined Euclidean shapes. In Greek and Roman architecture there was a logic and discipline that was rooted in Euclidean geometry when architectural form was derived primarily from symmetry and proportion. However, with the benefit of evaluating architecture through a 'fractal' lens it would be inconclusive to attribute this linear discipline to Gothic architecture, and the more natural architectural forms of later periods. Being characterised by self-similarity, the properties of fractal design are copies of itself at arbitrary scales, a paradigm for structure in the natural world, exemplified in objects as diverse as leaves, mountain ranges, clouds and galaxies (Sala 2007).

The word 'fractal' was coined by Mandelbrot (1977), and came originally from the Latin adjective *fractus*—from the verb *frangere*. These words resonated with the word fraction, and Mandebrot created the English and French word fractal, which despite its immeasurability in terms of length, nevertheless has a certain characteristic of roughness and describes the universal nature of complexity.

In Gothic architecture, working in a period of religious intensity, people of the highest intelligence and greatest ability developed strict geometric rules based on the understanding of natural forms, geometric principles and mathematical ratios thought to be the dominant ratios of the Universe. "Medieval theologians believed that God spoke through these forms and it is through such forms that they should appeal to him, thus Nature became the principal book that made the Absolute Truth visible. So, even when they applied the abstract Euclidian geometry, the Golden Mean and the proportional roots which they found in the proportions of living forms, governed their works" (Ramzy 2015).

The Age of Impressionism represented a departure from Euclidean geometry and a unique way of measuring 'physical reality' that heralded the search in art for more emotional paintings (Lorenzi and Mauro 2010). In a similar way, in *Fractals—The Patterns of Chaos*, Briggs (1992) explored human response to natural chaos and observed: "Most people find the haphazard profusions of Nature so intensely pleasing, even spiritually profound, that it seems common sense to say that there is an invigorating, even mystical order to the variable shapes of waves as they break, swallows on a summer evening, and the weather.

Mandelbrot (1977) observed in *The Fractal Geometry of Nature* that the "existence of these patterns challenges us to study forms that Euclid leaves aside as being 'formless' to investigate the morphology of the 'amorphous'. He argued that "in order to appreciate how the points, lines, planes and solids of the real world fill space, the Euclidian idea of distance (and measure) should be abandoned." Mathematicians have often disdained this challenge, however, and have increasingly chosen to flee from Nature by devising theories unrelated to anything we can see or feel.

Mandelbrot's new geometry of Nature, described by many of the irregular and fragmented patterns around us, has led to fully fledged theories, through identifying the family of shapes described as fractals. His discovery led to ways of calculating the fractional dimension of real objects, thereby determining a technique for constructing a shape and allowing geometry to make a claim on irregular patterns in Nature. This in turn informed the fractal dimension in architecture until then not fully comprehended or defined. He observed that the degree of irregularity remains constant over different scales, where over and over again a work would display a regular irregularity, a self-similarity. The fractional dimension is the degree of roughness or brokenness or irregularity of an object, such as a twisting coastline (Gleick 1997).

Mandelbrot's contemplation about 'linear fractals' was followed by a non-linear version that exists in a purely mathematical form. Contingent on this sequence of theories was "the third type of fractal [that] introduces a random element to the iteration and allows fractal artists to model the natural roughness or irregularity of waves, clouds, mountains, and so the fractal dimension of any complex, apparently chaotic feature, such as the branching patterns of a tree, can now be determined" (Briggs 1992). With an eye for pattern, the scientists who set the discipline of fractal geometry in motion "shared the taste for randomness and complexity, for jagged edges and sudden leaps" (Gleick 1997).

One definition of a fractal is that there is a structure in which there is substructure (i.e., complexity) at every level of magnification. "Magnifying a fractal by a fixed scaling factor, say 3, will give a set of pictures at magnification 1, 3, 9, 27, etc., all of which show structure and complexity. A "self-similar" fractal has the additional property that all these pictures are related by geometrical similarities—as long as one uses the scaling factor intrinsic to that fractal" further concluding that Modernism consists of 'pure platonic solids' that are incompatible with fractals, because the anti-fractal movement exists only on a single level (Salingaros 1999b).

"Books on architectural history emphasize how Modernist twentieth-century architecture is rational, being founded on mathematical principles" yet "the writings of the early modernists fail, however, to reveal any mathematical basis. Proposing pure geometric solids as 'mathematical' is totally simplistic" (Salingaros 1999b). The rectangular geometry of Classical architecture has generally informed the early Modern architects due to the dictates of styling as is evident where fractal properties are absent. This is notable in Mies van der Rohe's commitment to 'less is more', such as his German Pavilion at the Barcelona Exposition. It could be concluded that absence of the crucial characteristic of fractal subdivisions and scaling divides some Modernist work from buildings of all previous periods and styles.

In more recent contemporary architecture, self-similarity appears in assertive and intentional mode through the application of fractal geometry where irregular shapes occur in many different places and the property of self-similarity and the iterated function system is characterised by "Irregularity, self-similarity between the original structure and its smaller constitutive fragments, form invariance under changes of measure (scaling) and iteration of unit generators are the main properties which characterize fractal objects" (Sala 2007). Complexity and fractal geometry can inspire a sense of 'aesthetic measure', for which mathematician Birkhoff (1884–1944) proposed the definition: $M = OC$, is resonant (where M represents the 'aesthetic measure' (or 'beauty'), O the order, and C the complexity.

Yet for centuries scientists have dismissed such order. For a long time their attitude made sense. The traditional task of science has been to simplify Nature, expose its underlying logic as a means of control. But complex natural phenomena such as the weather cannot be stripped down, cleaned off and studied under a glass in a laboratory."

> Our feeling for beauty is inspired by the harmonious arrangement of order and disorder as it occurs in natural objects—in clouds, trees, mountain ranges, or snow crystals. The shapes of all these are dynamical processes jelled into physical forms, and particular combinations of order and disorder typical to them.
> Gleick (1997)

2.8 Thematic Transformation

Composers of music and music lovers understand that the transformations of a theme will always serve the purpose of 'unity within variety', the architectural role of the sonata form in the classical symphony.

Melody in music is a theme, musical phrasing that forms the basis for all or part of a composition. The development of a theme in music typically relies on repetition, making the melody assertive and to become the textural fabric, a sonic tapestry. The transformation of the theme, by small changes in a melody or cadence—without breaking the structure of the theme—is achieved by repeating the theme in an altered form involving harmony, counterpoint, rhythm, texture and application. These thematic attributes find their counterparts in the arts, and in architectural design.

> Kapustin uses many techniques to vary musical material and develop themes, including a theme repeated with varied scoring: a theme adapted for new key areas, making intervallic changes to accommodate the change in tonality; a motive is "spun out" in a free manner but with recurrences of the motive; and a motive is formally developed in the manner of a classical sonata
> Creighton (2016)

For the purposes of illustration, the well-known composition, the 3rd movement of Johannes Brahms' Piano Concerto No. 1 in D minor, Op. 15, conveys the essence of thematic transformation in music, where both orchestra and soloist are enlisted in the service of the musical ideas; technically difficult passages in the concerto are never gratuitous, but extend and develop the thematic material. As accessible to the ear is Frederik Chopin's Variations on "Là ci darem la mano" from Mozart's Don Giovanni, where the persistence of the thematic rendering is most striking.

By way of affirmation, thematic transformation is strongly evident in the Valse Oubliée No. 1 by Liszt where a theme is adapted for three different key areas, like a character in a play wearing three completely different costumes for three very different scenes. The first statement is in the tonic of F, the second in the mediant A minor, and the last in the unrelated key of G minor. Though the rhythm and melodic shape remain the same, intervallic relationships are modified to accommodate changes in tonality.

Similarly, in architecture, thematic transformation is holistically discernible in both the overall design and in the components constitute the whole, a unifying theme that individualises the subject, whether it be the melody in music, a design theme in architecture or its structural aesthetic.

2.9 Parametricism

In order to advance the discourse on structural design, it is important to explore the development of another widely recognised means, notably Parametricism, through which contemporary architects are able to design spatially and three-dimensionally in a manner never before enabled by the conventional approach to architectural design.

Parametricism, the new design paradigm, became currency after the 11th Biennale of Venice in 2008. Coined by Patrik Schumacher as a way of "addressing the demand for an increased level of articulated complexity by means of retooling its methods on the basis of parametric design systems. The contemporary architectural style that has achieved pervasive hegemony within the contemporary architectural avant-garde can best be understood as a research programme based upon the parametric paradigm. We propose to call this style: Parametricism" (Mentegazzi 2014).

Retrospectively, the traditional role of the master builder "began to change with increased sophistication of technologies in the nineteenth century, particularly in the second part of it. From mathematics, the theories of structures and applied mechanics were slowly developing to be used as specialised tools for predicting the behaviour of different types of structures and thereby refining the design to its optimum and most efficient level…those who specialised in these areas were to become civil engineers, and later to be even more specialised as structural engineers" (Khan 1980b).

For present day engagement with generative models from Nature, computational geometry is a branch of computer science that uses generative algorithms, thereby solving problems with complex geometrical aspects and outputs. Nevertheless, conventional design parameters familiar to the architectural fraternity remain the context and constraints of the site, the accommodation and facilities schedules, the building type and budget, and reliance on the designer's expertise for collective synthesis into architectural form. The algorithm acts as a set of rules that provides instructions to calculate and process data and carry out a specific task, following a predefined step-by-step procedure.

Form Finding is the search for optimal shape. In architectural design, form finding can be seen as shape finding with the optimum being rather subjective to the opinion of the architect. In structural design, the shape itself is not optimized, but the optimum is found in an optimal structural behaviour. For finding the optimal structural shape, different techniques can be used: analogical, digital and algorithmical.
Toussaint (2007).

Design software is available that combines script language with a visual interface, thereby providing the platform for a new variety of gridshell typologies (Mentegazzi 2014). In the application of this sophistication of the means does not alter the fact that: "Mathematics is a science of patterns. The mind perceives connections and interrelations between concepts and ideas, then links them together. The ability to create patterns is a consequence of our neural development in responding to our environment. Mathematical theories explain the relations among patterns that arise within ordered, logical structures. Patterns in the mind mimic patterns in Nature as well as man-made patterns, which is probably how human beings evolved so as to be able to do mathematics" (Salingaros 1999c). Mathematics in terms of modern applications has been taken to a significant new level in the service of many spheres without exception.

Simultaneity and Movement were the descriptors that Gideon in 1967 chose to explain his philosophy on developments in art and architecture. Giedion recognised the strong connection between art and architecture and argued that in Renaissance art the step is made to a third dimension by using the perspective, and a fourth dimension is added in cubism, the dimension 'time'. In architecture he described the fourth dimension as being reached by "a new transparency and interpenetration of space by the use of hitherto unknown technology and materials. The focus on the perspective is abandoned and replaced by this transparency" (Wesdijk 2012). These early observations regarding the new transparency and interpenetration of space could be construed as being prophetic, opening a vista on the contemporary generation of electronic aids and three-dimensional form-finding techniques.

Computer-aided design (CAD) has gained considerable traction amongst contemporary architects, and parametricism is the means through which architects are enabled to produce 'spatially' fluid designs. As a precursor to understanding parametricism, generative algorithms utilize computational geometry to produce design products providing a useful set of rules and instructions in a systematic procedure in order to process data to execute a defined task. The platform to perform such design processes uses Computer Aided Design (CAD) software, currently known as Algorithmic, Parametric or Generative design software, that should be deployed in the early design stage. Using these methods to deal with the physical aspects of buildings, some commentators refer to a new type of engineering emerging as 'bio-engineering'. (Khabazi 2012b).

The use of computer-empowering design tools opens the way for a global shift from formulaic and rigid geometrical constraint to enabling the continued evolution of architectural design to becoming increasingly complex and fluid, yet with 'fitness for purpose'. The reverse situation would trend away from the all-encompassing role of an architecture that responds to its purpose: utilitarian needs, contextual parameters, sustainability and energy efficiency. Logically, the intelligence that is required to invent and think through such correlations should precede computational implementation.

2.10 Future Paradigms

Writing about five ways that architects are redefining craftsmanship for a post-digital Age, Link (2016) observes that "traditional craft practices have been in decline since the industrial revolution, today a new generation of architects and designers has set about redefining and updating the notion of craft to include the most modern design and fabrication techniques around." In the publication *Autodesk's Line//Shape//Space publication* Link identifies some of the traits that connect architects and post-digital artisans to the craftsmen and women of a bygone era and observes that craftsmanship in the digital age is hard to define. He offers the prospect of the "age-old knowledge of woodcarvers, masons and other craftspeople" becoming "embedded in an intelligent design process using geometric computer models and machine fabrication to develop new crafts and architectural métiers."

2.10.1 Artificial Intelligence

The nature of intelligence has long puzzled philosophers and scientist, which might include skills, knowledge, alertness, wit or sympathy. Whereas computers can carry out tasks skilfully, store information and are responsive to commands, these attributes are only possible if digital technology is programmed to do so. Artificial Intelligence (AI), as we understand it on the other hand, would require a computer to address human abilities, such as reasoning, learning, judgement, remembering, emotion, intention, understanding, common sense and consciousness (Challoner 2002). The computer lacks these human sentient faculties, but is capable of creating 'virtual reality' situations which will transform the way design concepts are communicated and shared in future.

The introduction and speed of 3D printing into the toolkit means that physical models can be created at an early stage of a project, and "the model can be kept evolving as the project moves from conceptual through to detail design, leading to significant time and cost savings. The potential for this process to be extended into a tool for design and construction is already been demonstrated by architectural teams" (Ackerman 2016).

2.10.2 New Materials

At the turn of the twentieth century new ideas emerged in mathematics and physics that influenced new forms of Art development, namely Cubism, Art Nouveau and Modernism. (Lorenzi and Francaviglia 2010). In the twenty-first century, emerging from the materiality of Nano and Smart materials, architecture can be directed towards a systematic organism, where a material or construction can respond in action and reaction to its environment.

The development of Smart and Nano materials is bringing about a quantum shift in the capabilities of materials, potentially in current construction methods, and bring benefit from increased simplicity. Such advances in material processes require designers to keep pace and create standards through developing expert knowledge concerning the potential of their applications. A new kind of understanding and interdisciplinary cooperation is needed.

To explore the limits of these new technologies, the spectral transmissivity of Smart materials arises from their change and response to transient needs. Being photochromic materials, they change their colour when exposed to light: the more intense the incident light, the darker the surface. The use of Smart materials would thus find application in claddings, infill panels and the like but with limited structural potential. This is supported by the view of Addington and Schodeck (2005b) who observe that: "Not only are Smart materials a radical departure from the more normative materials in appearance, but their embedded technology has no precedent in the large integrated technological systems that are the standard in buildings." This argument is to some extent discounted by the opinion that Smart materials potentially contain several functional capabilities through which "construction methods will benefit from increased simplicity" (Daveiga and Ferreira 2005b). In terms of the aesthetical quality of a structure, at this stage the potential of Smart materials is yet to be demonstrated.

Dynamic research into biomimetic materials has begun through the growing interaction between the biological and material sciences. "Indeed, bio-inspiration does not result from the observation of natural structures alone, but requires a thorough investigation of structure-function relationships in biological materials. Nature has evolved a number of strategies to create outstanding functional properties with comparatively cheap base materials. This is achieved by hierarchical structuring, adaptive growth instead of fabrication, and constant remodelling and healing" (Frazl 2007b, c).

> The adoption of Smart materials poses yet another dilemma for the field of architecture. Whereas architects choose the materials for a building, engineers routinely select the technologies and design the systems. Smart materials are essentially material systems with embedded technological functions, many of which are quite sophisticated. Who, then, should make the decisions regarding their use?
> Addington and Schodeck (2005b)

2.10.3 Computer Aids

Comparing the design and construction methods of bygone eras and revolutionised patterns down the annals of history when new energy regimes brought about fundamental economic change, thus transforming the world order in general, this is generally described as the Third Industrial Revolution. The Fourth Industrial Revolution is the current trend of automation and data exchange in manufacturing technologies in what has also been termed a 'smart factory' environment, where cyber-physical systems communicate and cooperate with each other and with humans in real time. An integrated process with computer–aided design technology, with Smart material and revolutionary construction methods, will inevitably determine the future of how our buildings will be designed and given physical form. The implications of these technological advances for the design environment are potentially profound, given that before the advent of the digital Age, in all eras of architectural practice, the human hand draughted on paper the design and technical requirements for the complete construction of a building.

2.10.4 Cyberspace and Architecture

The appreciation of architectural design goes beyond the physical or virtual, it touches compellingly on psychic associations and somatic aspects. Conceptual links between architecture and cyberspace continue to transform our perception and the way space is conceived. A new cultural experience is being shared across a world-wide platform that awakens the potential for architecture to be navigated into a new realm of design expression emanating from the cyberspatial model, namely, the non-discernible influence that cyberspace facilitates in architectural form and design language, where discernibly the physical will be influenced by the architecture of the virtual. "This new type of visionary architecture will continue to stress the limits of what is possible encouraging the achievement of what is thought to be impossible" (Magermans 2004).

Firstly, the possibilities of architectural design in cyberspace requires an understanding of digital technology and the different tendencies that arise from three-dimensional calculation. An era of complexity can be predicted in both the education and in the practice of architecture, where the architects will need to see visual computation of architecture for its potential in their designs, and not only for its visual attributes. In order to achieve a reciprocity between the spatial grammar of virtual architectures with the form, function and symbolic content of physical architectures, the establishment of a framework would be necessitated to find the analogies.

> "No industry or enterprise is untouched by the persuasive influence of the information revolution. Understanding this revolution requires an examination of the determinants and sources of the value of information and the impact of that value on the organisational infrastructure of business and commerce"
> (Whittle 1996).

The interface between cyberspace and architecture is in the domain of information visualisation where the former should be seen as a primary informational tool. "Many authors are convinced that this 'space' possesses some valid rules and principles that should be followed by its designers. However, few of them dare to theorise these rules and write them down. In this respect, the principles Benedikt in *Cyberspace: Some Proposals in Cyberspace: First Steps* (1991) proposes are more than challenging, especially from the point of view that the future of architecture will be at least partially built in the virtual realm.

Today, the average person is experiencing architecture through an empowering lens. The advent of the smartphone has spawned the communication revolution that currently reaches over 1.75 billion subscribers. Predictably and based on current trends, this revolution will influence how the average person views the world and experiences architecture with unprecedented immediacy. The liberating outcome of the social media has enabled widespread public engagement and potentially the broader sharing of opinion about architecture as never before. Evaluation, choice and preference in architecture and the built environment has been placed in the hands of the people and been removed from the exclusive domain of experts and critics globally. Being an inevitable factor in the future paradigm shift, architectural practitioners and design trends are bound to respond accordingly (Kushner 2015).

References

References 2.1

Agostinho, F. S. (2005). *Architecture as drawing, perception and cognition* (p. 89). Background for an exercise of computer modeling applied to the Church of Sta. Maria de Belém – Lisboa.
Atalay, M. (2007). *Kant's aesthetic theory: Subjectivity vs. universal validity* (pp. 44–52). Stanford, CA: Stanford University.
Bachelard, G. 1994. *The poetics of space* (Vol. xxii). Boston: Beacon Press (in French 1958).
Bakhtin, M. M. (1990). *Art and Answerability: The Early Essays of M M Bakhtin*. In V. Liapunov, K. Brosteom, M. Holquist. (Trans.) (pp. 278–279). Austin, TX: Vadim Liapunov and Kenneth Brostrom.
Balkin, J. M. (2002a). *Review of cultural software – A theory of ideology*. New Haven, CT: Yale University Press.
Birch, R., & Sinclair, B. R. (2013a). *Spirituality in place: building connections between architecture, design, and spiritual experience* (p. 80). Alberta: University of Calgary.
Birkhoff, G. D. (1933a). *Aesthetic measure* (pp. 3–4). Cambridge, MA: Harvard University Press.
Cairns, G. (2012a, September). Architecture as political image: The perspective of advertising. *Architecture_Media_Politics_Society*, *1*(1), 1–13.
Ching, F. D. K. (1979). *Architecture: Form, space and order* (p. 300). New York: Van Nostrand Reinhold.
Ciftcioglu, O., & Bitterman, M. S. (2013). *Fusion of perceptions in architectural design*. eCAADe 2013. In Conference (Education and research in computer aided architectural design in europe) September 18–20, Delft, The Netherlands.
Croce, B. (2013). Judgement, criticism and taste. In. *Section 7, Croce's aesthetics* (*Revised version*). Stanford Encyclopedia of Philosophy.
Danielsson, C. B. (2011). *Experiencing architecture – exploring the soul of the eye* (p. 95, 101). Stockholm: The Royal Institute of Technology (KTH).
Dawkins, R. (2006a). *The selfish gene* (3rd Revised ed.). Oxford: Oxford University Press.
Doczi, G. (1985). *Proportional harmonies in architecture and nature: The power of limits* (pp. 3–16). Boston, MA: Shambala.
George, A. (2005). Study of architectural symbolism. In *Chapter 2: Development of symbolic pedagogical tools for communication in architecture* (pp. 24–25). Calicut: Department of Mechanical Engineering, NIT, University of Calicut.
Khabazi, Z. (2011). *Generative algorithms (using Grasshopper)* (p. 7). Digital version published by: Morphogenesism.
Khan, F. R. (1980a). *Structural aesthetics in architecture and its social and technological relevance* (p. 137). Article in Zeitschrift: IABSE congress report AIPC=IVBH, Kongressbericht Nov 1980.
Knight, R. P. (1972). *Analytical enquiry into the principles of taste* (p. 197). Gregg International, University of Virginia.
Leder, H., Belke, B., Oeberst, A., & August, D. (2004, November). A model of aesthetic appreciation and aesthetic judgments. *British Journal of Psychology, 95*(4), 489–508. doi:10.1348/0007126042369811.
Mitias, M. H. M. (1977). *The aesthetic object: Critical studies*. Washington: University Press of America.
Nasar, J. L. (1994). Urban design aesthetics: The evaluative qualities of building exteriors. In *Environment and behaviour* (pp. 377–401). Columbus: Ohio State University.
O'Byrne, K. K. (2009). Aesthetic appreciation. In S. J. Lopez (Ed.), *The encyclopedia of positive psychology* (Vol. 1 A, p. 21). Chichester: Wiley and Blackwell.
Peterson, C., & Seligman, M. E. P. (2011). Aesthetic appreciation. In S. J. Lopez (Ed.), *The encyclopedia of positive psychology*. Chichester: Wiley and Blackwell.

Pipes, A. (2003). *Foundations of art and design* (pp. 194, 198, 211). London: Laurence King.
Ramachandran, V. S., & Hirstein, W. (1999). The science of art. A neurological theory of aesthetic experience. *Journal of Consciousness Studies, 6*(6–7), 31–55.
Read, H. (1984). *The meaning of art* (pp. 17–48). London: Faber and Faber.
Salingaros, N. A. (1999a). Architecture, patterns and mathematics. *Nexus Network Journal on Line, 1*(2), 75–85.
Vali, A. P., & Nasekhiyan, S. (2014a). The concept and sense of place in architecture from phenomenological approach. *Indian Journal of Fundamental and Applied Life Sciences, 4*, 3746–3753.
Weber, R. (2008). The myth of meaningful forms. Philosophy and architecture. In W. T. Baker (Ed.), *Architecture of excellence in diverse world culture* (p. 68). Mulgrave: Images Publishing Group.

Online

Wythe, J. H. (2016). Accessed October 26, 2016, from http://www.alternative-architect.com/soul.htm
Neuroesthetics. (2016). Accessed October 1, 2016, from https://en.wikipedia.org/wiki/Neuroesthetics/
Khan, F. R. (2016). Accessed September 12, 2016, from http://www.e-periodica.ch/digbib/view?var=true&pid=bse-cr-001:1980:11::77#766
Vali, A. P., & Nasekhiyan, S. (2014b). The concept and sense of place in architecture from phenomenological approach, 4(S4), 3746–3753/Vali and Nasekhiyan Research Article © Copyright 2014|Centre for Info Bio Technology (CIBTech). Accessed October 9, 2016, from www.cibtech.org/sp.ed/jls/2014/04/jls.htm

References 2.2

Çizgen, G. (2012). What is context, why it is important? In *Rethinking the role of context and contextualism in architecture and design* (pp. 9, 12, 14, 30, 39). Submitted to the Institute of Graduate Studies and Research: Degree of Msc. in Architecture. Eastern Mediterranean University, Gazimağusa, North Cyprus.
Johnson, P. A. (1994). *Theory architecture concepts principles: Concepts, themes and practices* (pp. 39–284). Hoboken, NJ: Wiley.
Vali, A. P., & Nasekhiyan, S. (2014c). The concept and sense of place in architecture from phenomenological approach. *International Journal of Fundamental and Applied Life Sciences, 4*, 3746–3753.

References 2.3

Cairns, G. (2012b, September). Architecture as political image: The perspective of advertising. *Architecture_Media_Politics_Society, 1*(1), 1–13.
Dameron, G. W. (1991a). *Episcopal power and florentine society 1000–1320*. Cambridge MA: Harvard University Press.

Fletcher, B. (1948a). Gothic architecture 2: Architectural character. In *A history of architecture on the comparative method* (14th ed., p. 340). London: B.T. Batsford Ltd..

Coulton, G. G. (1928). *Life in the middle ages* (Vol. 1, 2nd ed.p. 24). Cambridge, MA: Cambridge University Press.

Frisch, T. G. (1987). *Gothic Art 1140-1450*C (p. 24). Toronto: University of Toronto Press, London in Assoc. with the Medieval Academy of America.

Gell, A. (1992). The technology of enchantment and the enchantment of technology. In J. Coote & A. Shelton (Eds.), *Anthropology, art, and aesthetics* (pp. 40–66). Oxford: Clarendon.

Marx, L. (1964). *The machine in the garden: Technology and the pastoral ideal in America* (p. 195). Oxford: Oxford University Press.

Milne, R. J. W. (1997a). *Structural engineering: History and development* (pp. 42–56). London: CRC Press. T&G Group.

Sudjic, D. (2006). *The edifice complex: How the rich and powerful shape the world* (p. 32). London: Penguin.

Online

Machiavelli, N. (2015). Accessed September 15, 2015, from http://www.nga.gov/content/ngaweb/features/slideshows/patrons-and-artists-in-late-15th-century-florence.html

Gaudí, A., & Güell, E. (2016). Accessed October 2, 2016, from http://www.gaudiclub.com/ingles/i_vida/relacio2.htm

References 2.4

Birch, R., & Sinclair, B. R. (2013b). *Spirituality in place: Building connections between architecture, design, and spiritual experience* (pp. 80–81). Alberta: University of Calgary.

Fletcher, B. (1948b). *A history of architecture on the comparative method* (14th ed.p. 597). London: B.T. Batsford.

Groys, B. (2013a). Becoming revolutionary: On Kazimir Malevich. In *Booklet: Russian Red (J Wasiliew). 21.09 to 25.10.2013* (p. 6, 8, 9).

Milne, R. J. W. (1997b). *Structural engineering: History and development* (pp. 42, 47, 56). London: CRC Press. T&G Group.

Sinclair, B. (2011). *Contemplating the spiritual ethos within a holistic framework for design and planning*. In 12th Symposium on Sustainable Development: Theories, Strategies and Global Governance Systems 1.1 (p. 4), Germany.

Tietz, J. (1999). *The story of architecture of the 20th century* (pp. 19–20). Cologne: Könemann.

Online

Pollard, N. (2016). Accessed September 15, 2016, from http://www.bbc.co.uk/history/ancient/romans/roman_religion_gallery_05.shtml

Hadid, Z. (2016). Accessed September 10, 2016, from http://www.theguardian.com/artanddesign/2013/sep/22/zaha-hadid-dont-make-nice-little-buildings

Groys, B. (2013b). Accessed September 12, 2016, from http://www.e-flux.com/journal/becoming-revolutionary-on-kazimir-malevich

References 2.5

Addington, D. M., & Schodeck, D. L. (2005a). Smart materials and new technologies – For the architecture and design professions. In *Materials and architecture 1.1* (pp. 201–238). New York: Architectural Press (imprint of Elsevier).
Daveiga, J., & Ferreira, P.. (2005a). *Smart and nano materials in architecture*. ACADIA05: Smart architecture. Austin: University of Texas.
Nervi, P. L. (1955). Concrete and structural form. *The Architect and Building News, 208*(27), 523–529.
Van der Merwe, D. (2013, August). Concrete, the liquid stone of the 21st century. In *Leading architecture and design*.
Wayman, E. (2011a). The secrets of ancient Rome's buildings. What is it about Roman concrete that keeps the Pantheon and the Colosseum still standing? Assist. Ed. Smithsonian Magazine.

Online

Wayman, E. (2011b). Accessed September 1, 2016, from http://www.smithsonianmag.com/history/the-secrets-of-ancient-romes-buildings-234992/

References 2.6

Ahuja, A. (2016). *Integration of nature and technology for smart cities. 13.1 Biophilic design* (3rd ed.pp. 331–332). Cham: Springer.
Arslan, S., & Sorguc, A. G. (2004). *Similarities between "structures in nature" and "man-made structures": Biomimesis in architecture*. In Collins, M W., & Brebbia, C. A. (Eds.), *Design and Nature II* (p. 48). Department of Architecture, Middle East Technical University, Turkey. WIT Press.
Benyus, J. M. (1998). *Biomimicry: Innovation inspired by nature*. New York: Perennial.
Benyus, J. M. (2002). *Biomimicry: Innovation inspired by nature*. New York, NY: William Morrow.
Carter, A., Kirkegaard, P. H., & Tyrell, R. (2013). The nature of tectonic architecture and structural design. In P. J. da Sousa Cruz (Ed.), *Structures and architecture: New concepts, applications and challenges* (p. 237). London: Taylor and Francis Group.
Frazl, P. (2007a, September). Biomimetic materials research: What can we really learn from Nature's structural materials? *Journal of The Royal Society Interface, 4*(15), 637–642.
Kellert, S. R., & Wilson, E. O. (1993). *The biophilia hypothesis*. Washington, DC: Island Press.
Root-Bernstein, R. M. (2001). *Sparks of genius. The thirteen thinking tools of the world's most creative people* (p. 145). Boston, MA: Mariner Books.
Weston, R. (2002). *Utzon: Inspiration vision architecture* (p. 11). Copenhagen: Hellerup Edition Bløndal.

Online

Søberg, M. (2016). Accessed November 12, 2016, from http://www.arcspace.com/features/zaha-hadid-architects/

References 2.7

Briggs, J. (1992). *Fractals, the patterns of chaos, A new aesthetic of art, science, and nature* (pp. 14, 69–70). New York: Simon & Schuster.
Gleick, J. (1997). *Chaos: Making a new science* (pp. 100–103, 117). London: Vintage.
Lorenzi, M. G., & Mauro, F. (2010). Art and mathematics in Antoni Gaudi's architecture: 'La Sagrada Familia'. *Journal of Applied Mathematics (Aplimat), 3*(1), 128.
Ramzy, N. S. (2015). The dual language of geometry in Gothic architecture: The symbolic message of euclidian geometry versus the visual dialogue of fractal geometry. *Journal of Medieval Art and Architecture, 3*(2), 138, 157, 165, 168.
Mandelbrot, B. (1977). *The fractal geometry of nature. Theme 1* (p. 1). San Francisco: W H Freeman.
Sala, N. (2007). Complex and fractal components in industrial design. *International Journal of Design and Nature, 1*(2), 161–173.
Salingaros, N. A. (1999b). Architecture, patterns and mathematics. *Nexus Network Journal, 1*(2), 75–85.

References 2.8

Online

Music Theory. (2016). Accessed November 14, 2016, from http://www.rhythmic.ca/music-tutorials/music-theory/melody/methods-of-theme-development.html
Classical Techniques. (2016). Accessed November 15, 2016, from https://www.scribd.com/document/61613927/Ch6-ClassicalTechniques
Creighton, R. (2016). Accessed November 16, 2016, from https://randycreighton.com/music/kapustin/Ch6_ClassicalTechniques

References 2.9

Khabazi, Z. (2012b). *Generative algorithms (using Grasshopper)* (p. 7). Digital version published by Morphogenesism.
Khan, F. R. (1980b). Structural aesthetics in architecture and its social and technological relevance. In *Zeitschrift: IABSE congress report AIPC=IVBH, Kongressbericht November* (p. 136).

Mentegazzi, E. (2014). In Mentegazzi, E. M. A. (Ed.), *Parametricism: The act of change in architecture* (p. 17, 38). Department of the Build Enviroment. Eindhoven: AnArchi.
Salingaros, N. A. (1999c). Architecture, patterns and mathematics. *Nexus Network Journal, 1*(2), 75–85.
Toussaint, M. H. (2007). *A design tool for timber gridshells* (p. 69). MSc Thesis, Delft University of Technology.
Wesdijk, M. (2012). Modernism and meaning. In *Siegfried Giedion's interpretation of modernism explained by his 'new space conception'* (pp. 1–3). Eindhoven: Eindhoven University of Technology.

References 2.10

Addington, D. M., & Schodeck, D. L. (2005b). Smart materials and new technologies – For the architecture and design professions. In *Materials and architecture 1.1* (p. 12). New York: Architectural Press (imprint of Elsevier).
Ackerman, D. (2016, March/April). *In three dimensions* (p. 25, 27). Johannesburg: Architect & Specificator/Promech.
Benedikt, M. (1991). *Cyberspace: Some proposals in cyberspace: First steps.* In M. Benedikt (Ed.), (pp. 119–224). London: MIT Press.
Birch, R., & Sinclair B. R. (2013d). *Spirituality in place: Building connections between architecture, design, and spiritual experience* (pp. 80–81, 86). Alberta: University of Calgary.
Challoner, J. (2002). In J. Gribbon (Ed.), *Artifical intelligence, Series: DK essential science, a beginners gide to robotics and man-made minds.* London: Dorling Kindersley.
Daveiga, J., & Ferreira, P. (2005b). *Smart and nano materials in architecture.* ACADIA05: Smart architecture. Austin: University of Texas.
Frazl, P. (2007b). Biomimetic materials research: What can we really learn from nature's structural materials? *Journal of The Royal Society Interface, 4*(15), 637–642. doi:10.1098/rsif.2007.0218.
Kushner, M. (2015). *The future of architecture in 100 buildings* (pp. viii–viix). New York: TED Books, Simon & Schuster.
Link, J. (2016, February 11). *5 ways architects are redefining craftsmanship for a postdigital age.* Autodesk's Redshift Publication.
Lorenzi, M. G., & Francaviglia, M. F. (2010). Art and mathematics in Antoni Gaudi's architecture: "La Sagrada Familia". *Journal of Applied Mathematics (Aplimat), 3*(1), 128.
Magermans, A. (2004). Architecture in cyberspace. In *Intelligent Agent 4.3.1.* (pp. 1–4).
Marques, S. O., & Goulette, J.-P. (2000). Architecture and cyberspace: Reciprocal spatial contamination. In *Construction the digital space. 4: SIGraDI,* Rio de Janeira (p. 66,67).
Mitchell, W. (1995). *City of bits: Space, place, and the infobahn* (p. 105). Cambridge, MA: MIT Press.
Van de Moere, A., Neuckermans, H., & Heylighen, A. (1998). *An architectural approach to cyberspace.* Belgium: KU Leuven University.
Whittle, D. B. (1996). *Cyberspace: The human dimension* (p. 306). New York: W.H. Freeman.

Online

Frazl, P. (2007c). Accessed September 12, 2016, from http://rsif.royalsocietypublishing.org/content/4/15/637. Accessed 12 September 2016.

Chapter 3
The Masters and Their Structures

Abstract In this Chapter a synthesis of the elemental determinants described in Chap. 2 is applied as a background to the ten chosen architectural works, whether through stylistic or philosophical genesis, and forms the discourse in this chapter. A review of the predominance of the 'structural aesthetic' in vernacular architecture confirms that historically the adoption of the equivalent elemental determinants that inform architecture of today existed albeit intuitively. Against the background of architectural monuments of the Greeks and Romans, the work of Filippo Brunelleschi is a fitting opening to the review, based on his stand-alone achievements as a precursor to the Renaissance. Since the Renaissance adopted Classical architecture for inspiration, architecture tended not to expose, but rather to clad the structure. In the fullness of time Antoni Gaudi (La Sagrada Familia, Barcelona) is identified as a major figure in returning architecture to its roots by adopting natural forms for his signature style in aesthetics. The figure of Le Corbusier (Ronchamp Chapel) looms large in architecture, and by virtue of his adoption of organic forms, left raw and unfinished, his gesture to the aesthetic of the structure is recognised. The review follows three towering figures during the middle of the twentieth century: Pier Luigi Nervi (Turin Exhibition Halls) who succeeded in promoting concrete in structural forms that could satisfy all the elemental determinants with impressive aesthetic outcomes; Oscar Niemeyer (National Congress Building, Brasilia) employed gravity-defying shell dome constructions that became the hallmark of his work in both Brazil and abroad; and Jörn Utzon's whimsical aesthetic that used spherical geometry in Nature for the Sydney Opera House in the most effective way to contain the greatest volume with the least surface area. Moving away from concrete, Frei Otto (Munich Olympic Stadium, 1972) forever changed attitudes to tensile structures that were unmatched in simplicity and elegance and the essence of good architecture. To bring the review up to present day, the avant-garde work of Zaha Hadid (Heydar Cultural Centre) opens a vista on the future, through the use of the most advanced digital design aids, where a steel spaceframe is used with fluidity, with a result bordering on the surreal in architecture.

Seen from a more secularised world, today the art of architectural design owes its freedom of expression through: a greater knowledge of scientific algorithms that engage: mathematics; the chaos theory (fractals); biomimetics; thematic transformation as found in both architecture and music; and parametricism in freeform

D. Thomas, *Masters of the Structural Aesthetic*, DOI 10.1007/978-981-10-5445-7_3

structures. In all cases the design approach to structure in particular has contributed to striking typologies where the aesthetic opportunities appear to be limitless. However, a cognitive evaluation of the structural aesthetic should not exclude recognition for significant examples in the wider global context than just the Western tradition.

Historically, the master builders of Africa and Asia employed the limited materials available to them to construct architectural works that reached heights of excellence and sophistication. "The spirit of place and community is fully expressed in their buildings, whose subtle variations are adapted to the specific social and economic, geographical and climatic requirements. The intelligence and virtuosity of these traditions, however, have been brutally obscured and despised.... Influential voices called for 'progress at any price', which they claimed was embodied in the 'International Style' of architecture" (Dethier 1981).

In both Mali and Morocco earth architecture at the hands of the artisans strikingly exploited the structure to express the aesthetic that represented those countries' cultural proclivities.

In Mali the towns of Timbuktu and Segou are almost entirely mud-built towns; mud walls, mud floors and streets and yards, sun-crazed mud plastering fluted with finger impressions. Djenne (Djenne-Jeno) from 250 to 300 BC has been a growing city situated on the Bani river, the biggest tributary of the Niger, and is one of the rare World Heritage towns today with an urban and architectural unity entirely of mud brick. Djenne is famed for its Mosque, the largest mud brick structure in the world (Fig. 3.1). It is indeed a magnificent structure, and is maintained on an annual basis with re-plastering being done voluntarily as a community effort. The buildings are most commonly built of red river mud in 'banco' technique, the making and building of sun-dried bricks or mud patties tempered with cereal straw. Gum arabic, an exudate of the local species of *Acacia* added to the mud for strength and has been in common usage. Ceilings are constructed of indigenous hardwood with large thicker diameter logs used as cross-corner joists with the intervening square space filled in with sapling poles or the trimmed basal rachi of palm leaves. The perimeter walls with Sudanic style bullet-shaped mud pillars and projecting turrets resemble tropical termite mounds (Maguire 2008).

Amongst the different traditions of earth construction, two major processes predominate: the first is pisé de terre, the Latin name first used in Lyon, France in the sixteenth century. This method involved the use of vertical parallel shuttering usually 50 cm apart between which the prepared earth was rammed. Where higher walls were required, the shuttering would be lifted progressively thereby ensuring a bond between castings. The second most common method is adobe—an Arabic and Berber word brought by the Spaniards to the Americas that is now part of the contemporary English lexicon. In the case of adobe, the carefully selected earth is mixed with water and vegetable fibres, such as chopped straw, to form a cohesive mixture. Cast into blocks and invariably sun dried, the finished walls would require a form of rendering for weatherproofing, and to prevent erosion. Typical forms of protection are an overhanging roof as a protection from rain, and a stone plinth at the foot of the walls from runoff and to prevent rising damp.

Fig. 3.1 The Djenne Mosque in Mali, and a Kasbah in the Dades Valley, Morocco. Traditional structures where the resulting structural aesthetic transcends the method of construction of sun-dried bricks tempered with straw. The distinctive means of construction is expressed in strong regional vernaculars where practicalities, such as the wood cantilevered treads, facilitate the task of maintenance of the mud rendering to the facades. The tree-like minarets illustrate the methods of construction and maintenance. The Djenne Mosque is one of the largest mud brick structures in the world

In Morocco, the kasbah is a fortified complex, built with local materials perfectly adapted to the climate. The role of the architecture was primarily defence but it also reflected the political and social structures of groups. Majestic fortresses of red earth or ochre, the kasbahs convey a unique recognisable identity, with the upper parts of the kasbah often decorated with Berber inspired geometric design themes found in other art forms, such as jewellery and carpets, and in tattoos worn by the women. Monolithic zigzag patterns in relief and the hollow motifs and shapes that evoke geometry with magical associations. The interpretation of these patterns in the collective unconscious for generations has probably become diluted. This integration of patterning is significant in terms of the aesthetic of a monolithic structure.

Arguably, for buildings built of mud, rain would present the main problem, but instead it is the unrelenting sun that cracks and dries the mud, causing crumbling. Many of the buildings of the 'toucouleur' and 'marocaine' types have been restored, thanks to international funding mainly from the Netherlands.

In Asia traditional structures of red sandstone are exceptional in terms of the heights of structural aestheticism achieved. The Red Fort Complex in Delhi was built as the palace fort of Shahjahanabad—the new capital of the fifth Mughal Emperor of India, Shah Jahan. The planning and design of the Red Fort represents a culmination of architectural development initiated in 1526. Named for its massive enclosing walls of red sandstone, it is considered to represent the zenith of Mughal creativity which, under the Shah Jahan, brought Islamic prototypes to a new level of refinement. Each pavilion within the fort reveals architectural elements typical of Mughal building, a good example being the scalloped repetitive red sandstone arching of the Diwan-i-Aam—the Public Audience Hall—that reflects a fusion of Persian, Timurid and Hindu traditions (Fig. 3.2). The Red Fort still stands as a powerful symbol of India's sovereignty, and the flag of India is raised on the rampart of the Larhoi Gate of the Fort complex every year on 15th August, India's Independence Day.

Fig. 3.2 The Public Audiences Hall, Diwan-i-Aam, in the Red Fort, Delhi (ca. 1526). Traditional structure built entirely in red sandstone, the awe-inspiring character of the Hall belies the simple structural form of repetitive scalloped arches resting on a rhythm of sculpted columns of its creation. The use of the limited use of building material heightens the sense of monument, and permits an aesthetic which signified the wealth and power of the Mughals and their empire

Fig. 3.3 Cité de Carcassonne (1853), fortified city, France. Traditional structure restored by the theorist and architect, Eugène Viollet-le-Duc in 1853 was added to the UNESCO list of World Heritage Sites in 1997. Constructed in brick, Viollet-le-Duc believed that the outward appearance of a building should reflect its rational method of construction

The medieval fortified city known as Cité de Carcassonne, was restored in 1853 by the theorist and architect, Eugène Viollet-le-Duc, and was added to the UNESCO list of World Heritage Sites in 1997 (Fig. 3.3). The Cité is of interest in terms of its structural aesthetic which reflects the philosophy of Viollet-le-Duc whose architectural theory was largely based on finding the ideal forms for specific materials, and using these forms to create buildings. His writings are important as he believed that materials should be used 'honestly' and that the rational construction of the building should reflect in the outward appearance of a building. Viollet-le-Duc observed that Greek architecture was a model for the correspondence of the structure with the aesthetic, and praised the Greek temple for the rational representation of its construction.

Of earlier significance before the Greek and Roman eras, was the Egyptian period when the pyramids, great monumental mounds of stone, were constructed to secure immortality of the soul, but it is important to recognise that the massive triangulated pyramidical forms located in the vast desert landscapes are of a scale that define a powerful 'sense of place'. The introduction of sturdy close-spaced columns of the Egyptian temples carrying stone lintols and based on simple constructional principles, also characterise the period.

As a prelude to later periods of architecture, it is instructive to return to the influence of the Greek Hellenic period (700–146 BCE) that ushered in a new structural form of which the most recognised, the temple, relied on a high order of the column and lintol method of construction and introduced the 'Orders of Architecture', the column and entablature types: Doric, Ionian and Corinthian.

Greek buildings designed on one constructive principle are characterised by harmony and simplicity.

The Greek temple aesthetic emerges from not only the important elements but also from the composition of the structure as a whole (Fig. 3.4). The column with its base, its shaft grooved vertically and tapering to meet the horizontal element of sculpted components, that in turn carry the triangular pediment supporting the roof structure, abide by conventions. There is a logic and discipline rooted in geometry about the size and spacing of the elements and the deliberate adherence to proportions in the relationships between them (Hahn 2012a). Refinement of the proportions and optical correction was practised for long horizontal elements, such as stylobates, architraves and cornices that would appear to an observer to sag or drop in the middle and these were given slightly curved outlines. In the Parthenon, to correct the appearance of falling outwards, the columns were inclined slightly inwards towards the top, the outer edge columns having a subtle entasis (a slight convexity or swelling) and inclination of nearly 7 cm in the height of 10.5 m so that the angled axes of the eight columns if extended would meet more than 1.61 km above the ground (Fletcher 1948a). An added visual correction was that the shafts of the columns are thicker at the bottom and taper gradually as they rise. Since against the background of the sky, the columns on the outer edges would appear

Fig. 3.4 The Parthenon, Athens, Greece (ca. BC 490), recognised as the epitome of the Greek architectural style. In implementing the columnar and trabeate construction style, the design principles of harmony and simplicity in Greek buildings are epitomised in the Parthenon. To correct an optical illusion, and to avoid visual sag, the horizontal lines of stylobates, architraves and cornices of the typical Greek temple were formed with slightly convex outlines. Vertical features, such as the columns, were slightly inclined inwards towards the top to overcome the illusion of falling outwards. The Doric, Ionic and Corinthian Orders of architecture refined the structural simplicity whereby the aesthetic whole is a product of the parts. For setting the benchmark that defines the structural aesthetic, despite structural forms being limited, the Classical Greek example remains relevant for main stream contemporary architecture

thinner, accordingly they were made stouter than those seen against the background of the naos. These subtle visual corrections exemplify Greek architecture in the great Hellenic Period (700–146 BCE).

Returning to the main focus of this chapter, which concerns ten iconic works in Western architectural tradition, those of the Greek, Roman and Gothic masons, Filippo Brunelleschi, Antoni Gaudi, Le Corbusier, Pier Luigi Nervi, Oscar Niemeyer, Jörn Utzon, Frei Paul Otto, and Zaha Hadid, all of which have a common theme—a recurring structural aesthetic is evident, developed historically from the hand tool and manual methods of the craftsmen of Roman times and the Gothic Age to the scientific knowledge, electronic tools and technological methods of Modern times.

3.1 Roman Builders (2 BCE–4 CE): The Pantheon, Rome

Roman architecture (146 BCE–365 CE) borrowed liberally from the Greek architectural styles. However, "When in 146 BCE Greece became a Roman province, her architecture, as a national style, died with her independence; although, like her civilisation, it continued to influence Roman and all later art" (Fletcher 1948a). Perhaps one of the most enduring edifices that demonstrates significant structural achievement from the Roman period is the Pantheon in Rome—the Rotunda.

Erected (120–124 CE) by the Emperor Hadrian, the Pantheon in Rome is a circular space with a diameter and height each of 43.5 m. "Roman mathematics was limited to rudimentary arithmetic and practical geometry... Given the scale of the complexity of the public buildings and infrastructure that Roman engineers designed and executed, it is surprising that they did not appear to be interested in the potential value of either Greek geometry and trigonometry or the applied mathematics of Archimedes" (Hahn 2012a).

Simple of form, the Pantheon's crowning glory, the vaulted dome, confirms the Roman builders' advanced constructional ability and use of materials that span over and enclose a large space, a significant development from column and lintol construction. The Pantheon's compelling aesthetic is largely derived from its expansive semispherical domed structure, fronted by a gigantic portico emulating the architectural elements from the Greek period. With its massive cylindrical walls, the expansive hemispherical dome, the elegant portico and a classically ornamented interior, the power of Roman engineering is combined with the aesthetic of the Greek Orders.

For the construction of the typical Roman arch, a wooden centering held the keystone in place during construction. Acting in compression a stable transmission of the gravitational force needed to be contained. Horizontal thrusts required the outermost piers to be broader to provide the required resistance. Roman architecture also made extensive use of curves, and spanned space with arches and vaults. Concrete was known and used extensively by the Romans and permitted new systems of construction which were used with originality and with distinctive aesthetic effect.

Fig. 3.5 The Pantheon, Rome, Italy (AD 120–124). Interior of dome (*left*) and springing (*right*) over Classical drum. Roman walls were of both stone and concrete and required a facing thus differing from the homogeneous marble walls of Greek architecture. The extended use of vaulting inherited from the Etruscans became standardised as a structural system and extended into the development of the dome. The Pantheon's octastyle Corinthian portico forms the grand imposing entrance and is an unusual merger of geometric forms

The discovery of concrete was fundamental. Easily poured, moulded, and shaped, it set to achieve the strength and resilience of stone (Fig. 3.5). The Romans discovered that when a certain volcanic powder is mixed with lime, sand, fragments of stone and masonry, and subsequently with water, the mix hardens to a substance that has stonelike consistency. Before it hardens, the mix can be shaped and moulded. Vitruvius described the volcanic powder as "a kind of natural powder from which natural causes produce astonishing results" (Hahn 2012a). The use of concrete rivalled that of stone or brick largely because of its ease of construction and it was relatively inexpensive.

The deliberate proportion of the square adds to the sense of repose of the generous span of the dome. A marble podium "supports the 6.1 m thick circular concrete wall, faced externally with brick banded with a layer of tiles every 0.9 m, and lined internally with marble and porphyry" (Fletcher 1948a). The inner surface of the hemispherical dome weighing 5000 tons is coffered at five levels, with mouldings adjusted for when viewed from below. The structural thickness is only 100 mm thick around the opening at the summit and "it is estimated that if normal concrete had been used for the entire dome, the stresses in the dome would have been eighty percent greater than they are, however, lighter materials such as small pots or pieces of pumice were used instead of stones in the higher levels of the dome" (Bondono 2016).

It is thought that the dome comprises a series of arches to redistribute the thrust to straddle the recessed openings below. The 8.50 m eye or occulus of the Rotunda may have had symbolic meaning: "that worship on this temple of all the gods should take place in a building open to the vault of heaven" (Fletcher 1948a). The design of the dome of the Pantheon appears to have been emulated later in the calidarium of the Thermae of Caracalla in Rome (211–217 CE) which is now in ruins.

The spanning property of the Roman vault was never equalled until the introduction of steel in the nineteenth century (Hahn 2012a).

3.2 Gothic Masons (1300–1500): Chartres Cathedral, France

The essential character of the Gothic period, particularly at the outset, was the predominance of the architecture. All the other arts were determined by it. In Gothic buildings the visual aesthetic was one of immense vitality; it was spikily linear and restlessly active. Romanesque architects had begun to substitute elasticity and equilibrium for the inert stability practised by the Romans, whereas the Gothic architect further extended the application of these static laws (Fletcher 1948b). Informed by the scholasticism and mysticism of the Middle Ages, this reordering reflected the exalted religious intensity, the pathos, and the self-intoxication with logical formalism that were the essence of the medieval architecture. By employing geometry in design as a tool to structure ideas and aesthetic impulses, or perhaps to incorporate into this work a meaningful system of symbols, it was the internal logic of the building that mattered more than achieving beauty (Ramzy 2015). The Gothic style was the dominant structural and aesthetic mode in Europe for a period of up to 400 years.

In Europe the nave of the Gothic cathedral would rise typically to three storeys with the arcade, triforium and clerestory levels. Chartres was perhaps the first of the great churches to dispense with the gallery level—normally between the arcade and triforium that featured in many early Gothic cathedrals—so that designers were able to make the richly glazed arcade and clerestory levels larger. The interior darkness added to the enrichment with the filtered light from a greater window area with richly coloured glass. With the increase in the area of the windows, so accordingly the wall area was reduced, necessitating the flying buttress, an innovative structural system that resisted the lateral thrust resulting in Chartres from the nearly 34 m high stone vaults, higher and wider than any attempted before in France. The pier design in Chartres Cathedral, known as *pilier cantonné*, was to prove highly influential and subsequently featured in a number of other high Gothic churches.

The sequence of operations in the construction of the Gothic cathedral required maintaining equilibrium and progressively overcoming the force of gravity. Fitchen (1981) in his book *The construction of Gothic Cathedrals: A Study of Medieval Vault Erection* attempted to answer these kinds of questions. His scholarly book is still the main source of reference on Gothic building processes. He observed that the sequence of construction was as follows: the piers of the high vaults were raised, while the flying buttresses were built at the same time. Tie rods were introduced to resist the consequential outward thrust until the completion of the abutments. The construction of the roof added weight to stabilise the piers and included the introduction of tie rods above the vault to complement those at the springing. Of significance is that during the complex procedures required by the Gothic masons in the erection of their buildings, it is averred that foreign expertise was called in to discuss the progress of the works.

With the benefit of hindsight, those spiritually inspired masters who were responsible for the entire epoch of Gothic architecture added significantly to the realm of the structural aesthetic, without distinction being made between structural aesthetics and architectural aesthetics. Interestingly, in many languages, the term engineer included all who were engaged in the art and science of designing anything from bridges to buildings.

The cathedrals of the Medieval period across Europe combined a new method of construction with gravity-defying results using small stones that were held in

Fig. 3.6 Chartres Cathedral (1194–1260 CE). Broadly, the evolution of Gothic architecture in Western Europe followed the Greek (trabeated) and Etruscan (arcuated), followed by Roman (trabeated and arcuated) from which Romanesque became characterised by the round arch and the Gothic by the pointed arch. The lofty roofs of the naves and transepts of the Gothic cathedrals, such as Chartres cathedral, utilised stonework to its utmost in the form of vaulting,, the downward and outward thrusts being counteracted by arches carried above the aisle roofs to press against the nave walls. The outer line of massive buttresses weighted by pinnacles were designed to resist the outward forces and became the aesthetic marking of the Gothic style. Chartres Cathedral dominates the town, and is recognisable by virtue of its dissimilar towers

equilibrium while forming the ‘rib and panel’ vaults over lofty church naves (Fig. 3.6). Using only units of brick or stone, the vault construction was stabilised in position by the surrounding flying buttresses, weighted by pinnacles. The structural parts of the building thus ceased to be its solid walls, and became a stone skeleton. “The stability of the Gothic cathedral depends on the proper adjustment of thrust and counterthrust.” expressed structurally through “a skeleton of piers, buttresses, arches and ribbed vaulting, all held in equilibrium by the combination of oblique and vertical forces neutralising each other” (Fletcher 1948b).

> The Gothic master builder “was a man who was taught by one or more master builders of the generation before him and during his training had to demonstrate his own sensitivity in aesthetics, his thorough knowledge of various techniques and methods of construction, his deep understanding of social needs and economic limitations, and finally his capacity to direct and manage all phases of construction of a building. The structural engineer did not exist until very recent times. He was indeed one of many personalities of the historical master builder.”
> Khan (1980a)

The theologians and the master masons of the Gothic period reproduced the geometry and followed the proportions of Nature, which contains a myriad of fractal patterns and which were thought to be the “divine language of the universe. Euclidean geometry was used to symbolise the cosmic order of the Universe for the eye of Heaven, whereas fractal geometry was used intuitively to visualise this order for the eye of the worshippers” (Ramzy 2015).

The finest in Gothic cathedrals abound. It is useful to examine at least a few which demonstrated the continual advancement of the structural principles which the stonemasons employed (Fig. 3.7). Chartres Cathedral in France represents this evolving architectural typology but so also does the Milan Cathedral in Milan, Italy (Fig. 3.8).

The late fourteenth and early fifteenth centuries in England saw a further development of vault construction rising to unimagined heights, given the reliance on, and limitations of small building blocks with which the stonemasons had to work. The ‘stellar’ vaulting ultimately led to the fan, palm, or conoidal vaulting in which the rising ribs were formed on concave cones using mathematical intricacy equivalent in its genius to computations used in contemporary structural design.

Theorists, such as Salingaros (1999) and Alexander et al. (1977), believe that buildings must convey languages, among them are pattern language and form language, and that geometry was used in the Gothic style to reproduce forms and patterns that reflect this dual language. In doing so, master masons distinguished between the abstract overall proportional lines of plans and elevations (form language), and the organic ornamental patterns on walls, ceiling, openings, and pavements (pattern language). They suggest that Euclidian Geometry was employed for the, usually invisible, proportional or working lines of the former, which are perceived only by the designer himself or by analytical view of the drawings with

Fig. 3.7 Oxford Cathedral, Christchurch (1158–1180). At the zenith of Gothic architecture, the distinctive structural aesthetic of perpendicular vaulting graced numerous examples, the cathedral at Christ Church College, Oxford being one. In the thirteenth century the design of the vault followed the outline of inverted, four-sided concave pyramids later being simplified by introduction of 'fan' vaulting

Fig. 3.8 Milan Cathedral, Italy (AD 1385–1485). The exterior of Milan cathedral is a gleaming mass of white marble with lofty traceried windows, buttresses and pinnacles crowned with statues. Without compromise, the aspirant character of the architecture reflects the vigorous civic pride of rival cities at the time. Patterns occur in the infinitely complex texture of finer and yet finer scales. Described as having nuances of Germanic influence as many of the architects employed on the building of it were from north of the Alps

a mesh of imaginary lines 'to address the eye of the Divine'; while Fractal geometry, was used for the a 'visual dialogue' intended for the earthly eye of humans.

The Renaissance that followed the medieval period, brought a new architecture and a revival of old Roman architecture, including the Classic tradition of the Roman orders, and the more evolved form of the dome and the vault.

3.3 Filippo Brunelleschi (1377–1446): Santa Maria del Fiore, Florence, Italy

The Renaissance was not so much the revival of ancient learning as the revival of learning itself. When the human mind was awakened to wider fields of interest than that which the church and Medieval times afforded. The change was sudden and violent, the authors are well-known and each successive change, through which the new art passed, is associated with the names of certain men: from Filippo Brunelleschi to Michaelanglelo to Palladio amongst others. It was the sudden and deliberate abandonment of an existing style and the adoption for imitation of an unfamiliar art of which the very principles had long been forgotten and had to be recovered—the bygone art of ancient Rome. Brunelleschi was foremost in the revolution in architecture. His dome in the cathedral of Santa Maria del Fiore in Florence (1420–34 CE) has been described as "the work which gave birth to scientific engineering in the Renaissance. Placed in context of early fifteenth century construction expertise, it must be regarded as one of the greatest building achievements of all time" (Zunz 1997).

Brunelleschi was born in 1377 and after an unsuccessful attempt to make a notary of him, his father apprenticed him to a goldsmith, "that he might learn to draw". He was also a student of ancient languages. He soon surpassed his master and, influenced by Donatello, he began to practise sculpture. Vasari tells of Donatello being beaten in competition by Brunelleschi with the sculpture of crucifixes.

> Many men are created by Nature small in person and in features," observed Vasari, "who have a mind full of such greatness and a heart of such irresistible vehemence, that if they do not begin difficult—nay, almost impossible—undertakings, and bring them to completion to the marvel of all who behold them, they have never any peace in their lives; and whatsoever work chance puts into their hands, however lowly and base it may be, they give it value and nobility… so lofty in intellect that it can be truly said that he [Brunelleschi] was sent to us by Heaven in order to give new form to architecture, which had been out of mind for hundreds of years; for the men of those times had spent much treasure to no purpose, making buildings without order, with bad method, with sorry design, with most strange inventions, with most ungraceful grace, and with even worse ornament (Vasari 1511–1574).

Having competed unsuccessfully with Lorenzo Ghiberti for the Baptistery Gates, Brunelleschi decided to visit Rome. With Donatello he spent four years in Rome where he studied architecture and sculpture of the ancient classics. When Brunelleschi saw the ruins of Rome for the first time he seemed beside himself. Both friends set to work to study and measure, sparing neither the time or money. No ancient work in Rome escaped them and Brunelleschi with no domestic ties

Fig. 3.9 Santa Maria del Fiore, Florence (1420–1434). The sight of Santa Maria rising above the roofscape of Florence demonstrates how the visual experience of architecture extends beyond the building itself, a phenomenological engagement of significance in all architecture. The vertical sections of the dome are Gothic arches, however, the dome profited from Brunelleschi's studies of the vaults of Roman structures that pushed Brunelleschi to execute this extraordinary dome. The Cathedral of Florence signals the beginning of the new art of building and marks the transition of architecture from the Gothic Age to the Renaissance

gave himself entirely to study. Two great thoughts inspired him: to restore architecture to the fine ancient model, and to leave behind him a name that would live in man's memory with those of Cimabue and Giotto; and to be able to devise a plan for vaulting the central part of the Cathedral of Florence, the Santa Maria del Fiore (Fig. 3.9).

After spending four years in Rome, during which time he became acquainted with the language of form and of the Classics, he returned to Florence. Models of the design of the dome for the Cathedral of Florence were being submitted in competition. Brunelleschi said that the dome must be built without centering (shuttering), but the popular opinion was that centering was necessary. Eventually he was appointed master, but to his disgust Lorenzo Ghiberti was appointed as his associate.

The main body of the cathedral was designed by Arnolfo di Cambio and built around the old church of S. Reparata. It was the dome over the octagonal transept that was added by Brunelleschi and that elevated the cathedral to high status and recognition using methods that experts do not fully understand even today. The cathedral's grandeur and artistry outshone those of any other structure in early Renaissance Italy, elevating Florence over rival cities. Though many believed it was too big and ambitious to build, architect Brunelleschi succeeded.

The vault measuring 42.25 m in diameter was erected without scaffolding and built freely up to a height of 35.5 m above the springing. It was cupola-shaped and double-shelled like earlier constructions. Its great elevation, due to the curve of a pointed arch, is augmented by a supporting drum pierced with large circular

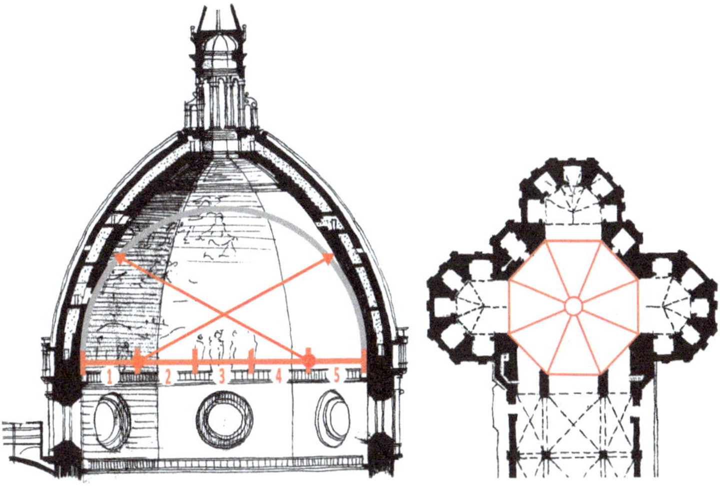

Fig. 3.10 The geometry of the dome, built on the octagonal base that is roughly 44 m wide, is complicated. It has inner and outer shells built to an elongated pointed profile. Not being a purely semi-circular dome, the focal point of the radius of the dome is centred one-fifth along the length of the base

openings—a first example of this disposition before it was often adopted later—but by first raising the springing of the vault, this did not facilitate the task.

The curve of Brunelleschi's dome is two-thirds of a quadrant, being raised acutely, since, as he explained, it carried the weight of the lantern, and would thereby diminish the overhang of the curve (Fig. 3.10). The weight of the brick panelling is distributed among the eight principle ribs and 16 intermediate ones. The inner shell of his double-shell design is about 2.28 m thick and the outer shell 0.75 m thick that together yield a structure that was far lighter and loftier than a solid dome of such size would have been. By using regular courses of herringbone brickwork into the texture of the cupola, little known before his time, the entire structure was given additional solidity (Mueller 2014a).

A strong wooden ring banded with iron was placed at the base of the dome to take the thrust. This octagonal composition neatly accommodates the nave and side aisles, in other words the whole width of the church, that provides striking individuality and distinction to the Cathedral amongst cathedrals.

The cornice and the small gallery to crown the drum on the outside were never completed, save on one face of the octagon, which faces south-east. Of the models for the lantern that were submitted in competition and in spite of all odds, Brunelleschi's design was accepted.

The boldness of the dome as an engineering feat is comparable to the work of the French engineer, Eiffel, whose bridges were constructed straight into space. By using inner and outer shells connected by solid brick arches, he effectively resisted the tendency of the dome to bulge outwards from the force of gravity at the springing of the dome. At this level a grid of interlocking stones hugs the dome like the hoop of a barrel. With the visual repose of the dome, and splendour of the entire composition, Brunelleschi's consummate art shows his mastery of the structural aesthetic.

Historical sources report that the dome was built without a temporary wooden formwork, or 'centering', to hold the masonry in place while drying. Brunelleschi himself left virtually no notes, so other details are uncertain. Many experts believe, for example, that ropes were used to mark the increasingly steep curved angle for laying the bricks, but they disagree about how that system was set up. It is possible to settle a method by which means the segments of the dome can be achieved, that is, by locating the centre of radii for each corner of the octagon at positions one-fifth along the cross-section at the springing, Fixing one end of the rope at these fulcra, and by stretching the other end to the required length, Brunelleschi was able to obtain a meeting point for the two circular arcs for each panel and form the junctions of the inner octagonal faces of the dome. Being level in a horizontal direction the masons were able to control the inner face of the octagonal dome as it progressed upwards.

Alberti, Michelozzi and other famous masters of the Age began to build churches and palaces in the style that Brunelleschi had developed. Work of such importance naturally must have permitted the originator a free rein for his genius in other work. Brunelleschi's work on the dome did not prevent him from attending to other great works, and he faced many difficulties—he was often called away with the result that his work was not entirely as he wanted it.

On March 25, 1436, the Feast of the Annunciation, Pope Eugenius IV and an assembly of cardinals and bishops consecrated the finished cathedral of Santa Maria del Fiore to the tolling of bells and cheering of proud Florentines. The lantern was added a decade later by another illustrious group who laid the cornerstone of the lantern, itself a decorative marble structure that Brunelleschi designed to crown his masterpiece.

Brunelleschi's religious structures are of considerable interest because of their elegant aesthetic, their proportions, and their variety and novelty of construction methods. In modern times, on the 567th anniversary of Brunelleschi's death, an architect from a Florentine family, Massimo Ricci, walked at the head of the memorial procession in Renaissance dress with "double ranks of halberdiers with gleaming weapons and breastplates, maidens in sweeping brocade gowns carrying wreaths of laurel and myrtle, drummers and trumpeters whose music filled the church and echoed in Brunelleschi's dome as on the day he was buried" (Ricci 2016).

Ricci spent much of his life trying to determine the construction techniques that Brunelleschi used to build the dome, as yet the largest masonry dome in the world. The most unsolved mystery is how his masons positioned each brick, stone beam

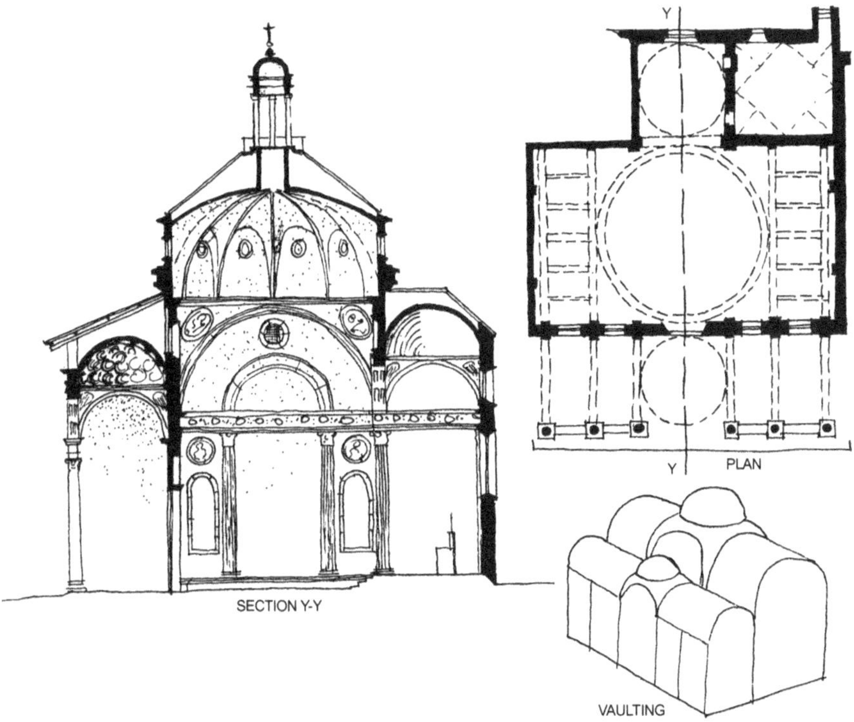

Fig. 3.11 Pazzi Chapel in the cloister of St Croce. The vaulting system used shows a full comprehension of Roman principles, as with the dome taking on a melon-like form. Certain members are differentiated tonally to suggest the sense of structure, that is to say the bones of the structure are expressed. The proportions show an understanding of the Roman language of form

and other structural elements with such precision over the vast space of the transept—modern builders without their laser levels, GPS positioning devices, and computer software would find Brunelleschi's structural method hard to match.

Hardly had he received the official commission to construct the dome of Santa Maria than he was asked to construct the mortuary chapel of the Pazzi Chapel in the cloister of St Croce (Fig. 3.11). Designed as a prostyle Roman temple with a façade of six columns, an ornate vault forms the frontispiece and vestibule leading to a square compartment covered by a dome on pendentives.

Brunelleschi was also commissioned to design the church of Santa Spirito, in Florence which was not completed until after his death. In its simplicity it follows basilican architecture, preferred by Italians at the time, and shows a freedom of method with the wide transepts that form a Latin cross with aisles around the nave, transepts and the choir. A flat ceiling covers the nave with a dome over the crossing with a series of little domes on pendentives.

Brunelleschi died on April 15, 1446. Although as an architect he was considered a humble craftsmean, he was honoured along with the saints and is buried in the crypt under the dome Santa Maria del Fiord which is his lasting legacy as a 'master of the structural aesthetic' A memorial plaque nearby celebrates his "divine intellect."

The Cathedral of Florence signals the beginning of the new art of building and marks the transition of architecture from the Gothic Age to the Renaissance (Hahn 2012b).

> The Wardens of Works of Santa Maria del Fiore and the Consuls of the Guild of Wool, assembling on his arrival, explained to Filippo all the difficulties, from the greatest to the smallest, which were being raised by the masters, who were in his presence at the audience together with them. Whereupon Filippo spoke these words: "My Lords the Wardens, there is no doubt that great enterprises ever present difficulties in their execution, and if any ever did so, this of yours presents them, and even greater than perchance you are aware of, for the reason that I do not know whether even the ancients ever raised a vault so tremendous as this will be; and although I have often pondered over the framework necessary both within and without, and how it may be possible to work at it securely, I have never been able to come to any resolution, and I am aghast no less at the breadth than at the height of the edifice, for the reason that, if it could be made round, we might use the method used by the Romans in raising the dome over the Pantheon in Rome, that is, the Rotunda, whereas here we must follow the eight sides, and bind the stones together with ties and by dove-tailing them, which will be something very difficult.
> (Vasari 2012)

3.4 Antoni Gaudi (1852–1926): La Sagrada Famila, Barcelona, Spain

Antoni Gaudí (Antoni Plàcid Guillem Gaudí i Cornet) was born in Reus, Catalonia (or maybe in Riudoms) on the 25th of June 1852. Gaudi's revolutionary projects were created in a climate of change when, at the turn of the twentieth century, mathematics and physics led to new forms of Art, such as Cubism, Art Nouveau and Modernism (Lorenzi and Francaviglia 2010).

After completion of his architectural studies at the age of 17 years, Gaudi's first buildings from 1878 onwards showed his interest in motifs from Islamic Models. Under the influence of the Catalan movement (Renaixenca) from an early age, followed by Modernisme around 1890, he was of the era when after several 100 years, political and artistic revolt was being articulated with Catalonia staking out its claim to independence from the rest of Spain.

The most famous of his work is the Basilica Church 'La Sagrada Familia', the neo-Gothic cathedral which he did not complete before he died—"a fantasmagoric constructive project, in which curved Geometry and Fractals dominate along with remnants of antique esoteric knowledge" (Lorenzi and Francaviglia 2010) (Fig. 3.12).

Gaudi's buildings are a "synthesis of structural integrity and sculptural flamboyance which places him as one of the great designers and innovators of the last century.... He rejected the fashionable Gothic revival of the late nineteenth

Fig. 3.12 La Sagrada Familia, Barcelona (commenced in 1882) In the Middle Ages Catalan was one of the group Provinçal dialects which are understood on both sides of the Pyrenees today. Awareness of the language accounts for a greater self-determination. Around 1900 Catalan was practically the only language spoken in Barcelona. It was against this background that architect Antoni Gaudi won the public hearts by including the familiar amidst the bizarre, gothicised constructions taking their place alongside shapes drawn from Nature and giving a sense of organic growth

century, a revival which attempted to replicate a 600 year old style without regard to contemporary expertise. But he used Gothic symbolism in a new and totally personal and idiosyncratic manner" (Milne 1997a). The creative force of his legacy lives on in the modern day initiatives to complete the cathedral that he started, following Gaudi's own principles of method and technique which bordered on genius.

The four towers reflect a (neo)-Gothic design and are reminiscent of honeycombing found in termitiaries... "These towers are terminated by cusps having a precise geometric form, covered by multicoloured ceramic tiles, certainly influenced by Cubism (they were in fact finished around 1920); their pinnacles are a composition of various intertwined geometrical elements" (Lorenzi and Francaviglia 2010).

Gaudi's deployment of the principles of Nature and mathematics are of no less importance. The continuous reference to Nature from which the structural elements are derived is given by the dynamical undulating contours, echoing the curves and shapes of trees and flowers, as well as the strong visual allusion to the bone structures that are evident in biological species. In the field of science a revolutionary idea of fractals was challenging the straight line and radial curve of Euclidean geometry, the "self-reproducing objects at smaller and smaller scales that at the same time show an ordered and a chaotic behaviour... On purpose or not, it is remarkable that Gaudi used a lot of fractal ideas in his Architecture" (Lorenzi and Francaviglia 2010).

On entering the crypt of La Sagrada Familia, the four inclined basalt columns stand out, and give the sensation of an organic and natural structure, like the trees in a forest. Taking the eucalyptus tree as his example, the network of columns in the main nave of cathedral resembles a forest executed in stone (Zerbst 2013).

Gaudi has been described as an outsider in terms of his individualism in architecture—not so. In using structure to express his architectural objectives he not only did what many Gothic masterbuilders and Renaissance builders did before him, but he also explored new forms that were intended to follow the thrust lines more directly (Fig. 3.13). By beginning to work in the Gothic architectural style (he was influenced by Viollet-le-Duc) his own distinctive architectural language developed.

To authenticate the origin of his structural forms in Nature, Gaudi employed catenary models involving ropes that that hung from the points where columns had to be placed. Small sacks of lead shot—weighing just a fraction of the weight the real arches would have to support—were hung from each arch formed by the ropes, hanging from points traced on wood and suspended from a ceiling matching the outline of the church (Lorenzi and Francaviglia 2010). Rotated into vertical structures, these catenaric arches were the eventual pressure lines of the structure and account for the natural forms in his work. In geometry, as well as physics, 'catenary' is a curved shape that is derived from a hanging chain or rope when fixed at its ends and a U-shape results, similar to a parabola, or the graph of a hyberbolic cosine.

Fig. 3.13 The four towers (*above left*), inherited from the neo-Gothic Architecture, are a reminder of termite mounds and terminate with cusps that have a precise geometric form, covered by multicoloured ceramic tiles. The image of a model (*top right*) depicts Gaudi's vision for the completed cathedral. The Passion Façade (*below*) is austere, plain and simple, with ample bare stone, and is carved with harsh straight lines to resemble the bones of a skeleton. Dedicated to the Passion of Christ, the suffering during the crucifixion, Gaudi intended the facade to strike fear into the onlooker and show the severity of Christ's sacrifice

Gaudì saw God manifest in Nature which resonates with "why Gaudì was searching the presence of 'beauty' in force lines and funicular polygonals, with the consequence of preferring helicoids and parabolas with respect to circles and squares." Gaudì was moreover a fervent Catholic "to the point that his "scientific principle" becomes in fact the counterpart of a "supranatural revelation". He once said: "Those who look for the laws of Nature as a support for their new works collaborate with God" (Lorenzi and Francaviglia 2010). *The General Morphology*

Fig. 3.14 Many decorations of the Basilica are directly referable to the Art Nouveau and, as such, to the geometric shapes that characterize this new artistic style of the beginning of the twentieth century. Antoni Gaudi's design language subtly combined the simple shapes of Geometry with the spirit and forms of Nature

of Organisms of Haeckel—translated in Spanish and published in Barcelona in 1887—would have been a source of inspiration for the fantastic geometric and natural shapes that Gaudì used in his work.

At the turn of the twentieth century, art form movements, such as Art Nouveau, evolved and put their stamp on all the arts, including architecture (Fig. 3.14). These new styles referenced their inspiration from Nature, which in turn liberated formulaic structural design through the use of dynamic and fluid flowing elements, trees being an obvious source for the design of vertical structural elements. In his profound use of natural forms, such as the tree with its natural fractal structure, Gaudi cleverly deployed Euclidean geometry in a way that was analogous to the fractal process of self-similarity and self-reproduction. Instead of relying directly on the "simple shapes" of geometry he achieved this by mimicking Nature in subtle combinations. In his work rotational surfaces having a peculiar role, such as hyperboloids and paraboloids that were in fact borrowed from Nature, so to allow his work to resemble environmental elements.

Gaudi's singular ability to embellish the fabric of his buildings with elements consistent with his Nature-inspired structural forms strengthens their aesthetic integrity: iteratively and in reverse, 'the whole becomes the sum of its parts'. However, it would not be wrong to explore thematic transformation, also known as thematic metamorphosis or thematic development in music favoured by Romantic music composers in which a *leitmotif*, or theme is changed by permutations of augmentation, diminution, and fragmentation.

Fig. 3.15 Casa Batlló (1906) (*left*) and Casa Milà (1906–1912) (*right*) reflect the height of Gaudi's success in his lifetime. Casa Batlló, better known as La Pedrera, is a demonstration of the curved surface, of an animated façade in stone and in curves that gradually pass from planal to ovoidal in shape. Casa Milà with its gently bowed façade and irredescent surface of ceramics is playfully alive

Thematic transformation is also strongly expressed in the facades of both Casa Milà, and Casa Batllö (1905–1907) in Barcelona, where the natural and organic forms are not of purely applied decoration, instead the structural columns at ground level are bone-shaped, echoing the natural component forms of a skeleton so that the thematic association with natural elements is unavoidable (Fig. 3.15). Gaudi's ability to resort to higher inspiration for his art is demonstrated in the house of the family Güell where a celestial theme in the central hall is a veritable reproduction of the Cosmos (Lorenzi and Francaviglia 2010).

Chronologically, the work of Antoni Gaudi comes first to mind as a individual creative force through which his architectural expression reached for the stars, figuratively and structurally. At the time it was conjectured whether he was a nut or a genius, Antoni Gaudi was nevertheless awarded the title of Architect in Barcelona in 1878. Time told that he was in fact a genius.

Gaudì died at the age of 73 years in Barcelona in 1926, leaving to the posterity his most famous and incomplete work of architecture: the Basilica Church of La Sagrada Família—a fantastic monument still under construction. "It has been argued that Gaudí—conscious of the fact the construction of the Church would much probably extend well beyond his death—preferred not to complete the perimeter of the Church but rather dictate its geometric "vertical" structure by completing some of the geometrised towers" (Lorenzi and Francaviglia 2010).

The Basilica Church is still under construction, and the towers and most of the church's structure are to be completed by 2026, the centennial of Gaudí's death; decorative elements should be complete by 2030 or 2032. Through the simulation of Gaudi's structural methods towards completion, the revolution in design that he left to the town of Barcelona and the architectural world his unique heritage will recognised.

3.5 Le Corbusier (1887–1965): Chapel of Notre Dame du Haut, Ronchamp, France

The Modernists, or the protagonists for the International Style, are associated with names such as Le Corbusier, those whose works became known by their individual interpretations, through their architectural philosophies, and as expressed in their buildings.

Fig. 3.16 Notre-Dame du Haut Ronchamp, Vosges (1950–1955). Not only providing for a pilgrimage for the ascent to this lofty site, the extraordinary architecture and ambience of its setting serves this meditative shrine well. Three elements dominate: on the left as one approaches, the towering silo-like chapel; the tapered edge of the protective concrete roof with its unadorned curved soffite that also projects over the outside chancel; and the heavy, prow-like and battered supporting wall with the deep openings and coloured glass insets. Dominated by the powerful presence, the roof recalls a Biblical desert tent. The plastic shapes flow and counterflow, hide and reveal an aesthetic that arises from the deliberate departure from angles and straight lines, achieved through a unique composition of the structural elements

Le Corbusier was born as Charles Edouard Jeanneret on October 6, 1887 in La Chaux de Fonds, Switzerland. He studied at the La Chaux de Fonds Art School, the commencement of a career spanning five decades. Of his most iconic buildings the Chapel of Notre-Dame du Haut in Ronchamp is the most striking departure from his earlier buildings where, in the Chapel, geometric shapes were from other sources of inspiration, and where the ode to the rectangle had given way to the sensual curve (Fig. 3.16).

Form that is associated with Algerian vernacular architecture is evident in the Chapel but, although not the initial inspiration for the architecture, the Chapel appears completely organic both in form and in materials that are left in the raw and allowed to age naturally, the plasticity of form giving the feel of sculpture. Danièle Pauly published extensive researches on Ronchamp, and observes that the design is a symptom of revolt against reason: "a rejection of the overly rigorous forms and the orthogonal system that had prevailed for decades" (Evans 2000). The complex shapes at Ronchamp start from a theme of acoustic parabolas, and the esoteric geometric shapes from Le Corbusier's earlier buildings have given way to more subtle, fractal, 'natural' shapes, leading to the description of Ronchamp by some commentators as the first Post-Modern building.

The dominant feature of the Ronchamp Chapel from the exterior is its roof which is a one-of-a-kind sculpture in concrete where the visual effect is so powerful that building's expressionistic language becomes meditative in its aesthetic, the smooth curve of the roof being likened to the symbol of 'praying hands' (Fig. 3.17). Le Corbusier described what he did at Ronchamp as "totally free architecture", a reflection that is underlined by Labatut who concluded that "Ronchamp expressed a liberation for Corbusier from Corbusier." The roof is constructed with two reinforced concrete membranes with soft curves lifting upward toward the sky in a dark neutral colour that contrasts greatly with the whitewashed walls below. There is a line of light entering through the small gap between the roof and walls of the Chapel so that seen from the darkened interior, the roof appears to be floating. The aesthetic of the roof is further heightened through allowing the linear texture of the timber formwork to emphasise lines of increasing curvature.

Surrealism is a key to the Ronchamp chapel, yet programmatically the church is simple—an oblong nave, two side entrances, an axial main altar, and three chapels beneath towers—as is its structure, with rough masonry walls faced with whitewashed Gunite (sprayed concrete) and a roof of contrasting beton brut, ruled or shuttered concrete surface, emphasising the lines of increasing curvature. Le Corbusier had insisted on the rationalisataion of construction and the extensive use of concrete for its purity and rendering, permitting the idea of 'brutalism' that describes the aesthetics of raw concrete. By using the indentations of planks of wood, the effect of plasticity has been created, with the veining and lines of the joints to underline the strength of the mass, and to accentuate the features of an element, such as the external pulpit. Although commonplace in civil engineering, such as in bridges, since the 1930s in modern architecture the ruled surface became

Fig. 3.17 The chapel towers (*top left* and *right*) rising above the roof line orient outwardly as if to engage spirtually with the wider world. The south-facing wall of windows (*right*) is also known as the 'wall of light'. The windows vary in size and are placed in no certain pattern adding colour to the otherwise chromatic interior

symbolic of progress in architectural expression and contemporaneity, a new direction that Le Corbusier used to write so much about.

Detracting from the 'poetic' statement that the Chapel makes, Evans (2000) proffers a somewhat mundane analogy for its towers that "are a bit like silos; its south and east walls a bit like the acoustic early warning dishes built of concrete in 1928 on the English coast to warn of enemy attacks." Formally and symbolically, however, this small building, which is sited atop a hillside with access from the south, is immensely powerful and complex" (KTStudioKT.net).

In his book on Ronchamp, Le Corbusier wrote that this building was "informed by all-embracing mathematics" and declared that the Modulor was everywhere. He defied the visitor to give off-hand the dimensions of the different parts of the building. Le Corbusier developed the Modulor system or proportion between 1943 and 1955, basing the system or proportion on the proportions of man and is described by Evans (2000): "Take a man-with-arm-upraised, 2.20 m in height; put him inside two squares, 1.10 × 1.10 m each, superimposed on each other; put a third square astride the first two squares. This third square should give you a solution. The place of the right angle should help you decide where to put this third square."

On entering one finds an interior totally sculptured in its shape and in its ability to enwrap and make a participant of the spectator. Here one finds no reference to known rooms; here, actually, are few right angles or straight lines. For this is a wonderfully plastic shape, flowing and counterflowing, hiding and revealing—all in an area of some 13 by 25 m. Dominated by the powerful underside of the roof, a soffite recalling a Biblical desert tent, and ablaze with a myriad of colours from myriads of irregularly placed, deepset windows on the south, it is punctuated by unexpected accents of the vertical side chapels...
Kidder Smith (1961)

Well-known Australian writer, academic and Guardian columnist, Germaine Greer took up the challenge and visited Ronchamp to establish Le Corbusier's claim that it represented the Modulor in scale and proportion. Greer (2016) observed: " Though everything about Ronchamp is ebullient, the Modulor principle is applied throughout, as Le Corbusier's drawings make clear. The apparently impossible roof, flapping clear of the walls like a nun's starched bonnet, is certainly a concrete version of dizzy rapture, but it is also comprehensible in anthropometric terms. The tapering walls that zoom endlessly upwards, dragging the eye up to the monumental formwork snagged on the topmost point, are as massive as they need to be and no more. The organising Modulor principle is what makes possible the simultaneous expression of order and energy."

Le Corbusier insisted the use of the Modulor at Ronchamp was fundamental, against massive evidence amongst the sceptics, recognised commentators, who had defined the metric dimensions of the Modulor.

The project was carried out by a handful of masons under the direction of André Maisonnier, architect from the region. He reported that "We had studied geometry a great deal, but at the extremities of the [roof] shell geometry did not enable us to find the junctions and express the double curvature. I had worked very hard at this enterprise. . . At the office there were always engineers who would calculate what was necessary." Where Le Corbusier himself likened the roof to a crab shell,

Maisonnier likened it to an aircraft wing and described the most characteristic geometric elements being the roof shell, "composed of two reversed conoids parallel to one another (these are the two concrete skins of the roof), and the south wall composed of two opposed rule surfaces that start obliquely from the main door, straightening up as they pass along the plan of the wall and arriving as two verticals at the southeast corner of the building" (Evans 2000).

Since its inception, Ronchamp has been a mecca for people who undertake the pilgrimage up the hill to its prominent site, where it stands palpably inviting. Greer concluded: "Fascist architecture is meant to make people feel small and insignificant; Le Corbusier's intention was the exact opposite, but until I went to Ronchamp I didn't get it."

Ronchamp continues to elude definition in terms of its place in architectural history. However, regardless of all scepticism the building should be seen as a singular architectural statement in the widest context that relies to a great extent on its 'structural aesthetic'.

3.6 Pier Luigi Nervi (1891–1979): Exhibition Hall B, Turin, Italy

Recognition for the work of Pier Luigi Nervi gathered momentum in the latter half of the twentieth century, when through his operations and achievements in physical form, he demonstrated his commitment and architectural schemata to structural honesty, structure without embellishment. Of significance is the machinery that Nervi brought into play whenever he undertook a project.

Nervi emerged in the 1950s as a versatile practitioner, bridging architecture and civil engineering, having come from the educational systems of Europe where a unified approach to the art of building was manifest (Fig. 3.18). The true genius of Nervi was his clear, honest, logical and yet imaginative structures and construction. He decried ideas of buildings which only followed the popular discourse and which could only fill up museums. Nervi admired and marvelled at the historical achievements of the Gothic builders, appreciating their integrity from the logical interpretation of how materials should be used, and their application to geometric forms that gave structural strength and art forms. "His explorations in geometry were bold and extremely imaginative yet always contained a strong, disciplined rigour that avoided illogical extravagance. In this regard it is helpful to remember his firm criticism of the design proposal for the Sydney Opera House" (Melaragno 1991)—see later.

Ernesto Rogers' *Portrait of Pier Luigi Nervi* encapsulates Nervi's approach to buildings which he proclaims are like a delicately woven fabric, a system of nerves made from structural elements purged of all frills so as to seem always in vibration. Their harmony is based on an equilibrium of tensions and of restrained, disciplined movement. This was widely accepted as also true of his personality. He saw more in

Fig. 3.18 Turin Exhibition Hall B (1947–1954). The nature and physical outcome from the relationship between aesthetics and technology—architecture and structural engineering—was the starting point for much of Pier Luigi Nervi's work. Nervi's philosophy is rooted in the conceptual, and a creative, pragmatic understanding of material properties and their visual and structural patterning. In his buildings, there exists a union of an ordinary material like concrete with form, structure and pattern, and his buildings show organic reference. Masterful at making the flow of forces visual, Nervi called to question the similarities between artist, architect and engineer. Being non-formulaic, his structural patterns convey clarity of mind and a sound logic of construction

objects than just shape, he saw the essence of things where true vitality can be found—this amounted to a psychological approach or intuitive sense to deduce the essential qualities of good design. His work reflects his strength and a confident personality.

In the work of Nervi a synthesis is evident between science and art, as well as use and means by reason of his classical approach. For medieval craftsmen this synthesis was unconscioius, and with the Renaissance architect, deliberate. In this context the words of Leon Battista Alberti are recalled: "Him I call Architect, who, by sure and wonderful Art and Method is able, both with thought and invention, to devise, and with execution, to complete all those Works, which by means of Movement of Great Weights, and amassment of Bodies, can, with the greatest Beauty be adapted to the uses of mankind." Further, "In that synthesis the artistic understanding of form and of its material content must not be a dual conception leading to divergent roads, but rather a dramatic process where both are interchangeable. When this process depends on two opposing forces, the architect has to resolve them into a synthesis" (Rogers 1957).

> The question arises about the bearing of Nervi's work on contemporary architectural thinking. The benefits in learning from Nervi are as relevant today, as many designers have a strong interest in the relationship of aesthetics and performance tempered by material properties, construction technology and economics. Accordingly, this team aims to further the tangible relationship of aesthetics, material deposition, and technology—relationships that Nervi masterfully engaged.
> Clifford (2011)

On the authority of Rogers, from his close association with Nervi, he states that Nervi was under the illusion that some of his most brilliant solutions had been reached solely by the disciplined and precise mind of an engineer, that the beauty of his work was due to a neutral or passive process of thought having nothing to do with aesthetics. If this was so then Nervi was an artist against his will, even sometimes against his own theoretical conviction. Nervi's special gift is an expression, through mathematical severity, of a perfect structure which, although a result of complicated formulae, appears so simple that it could have been observed from the laws of Nature rather than arrived at through human thought. Nothing is more difficult than to express complex things in a simple way. Nervi confesses that sometimes he succumbs to the temptation of striving for beauty which threatens to overshadow the simplicity of his structure, but then his belief becomes the moderator: "My belief in the inherent aesthetic force of a good structural solution was never shaken" (Rogers 1957).

'Beauty is the glory of Truth', is the basis of a philosophy which both Nervi and Mies van der Rohe preached. For the latter 'less is more' became the overriding mantra for Truth and was expressed differently in his architecture.

Nervi always reminded others of the limitations and dangers in the training of engineers where insufficient account is taken of empirical knowledge and of the "feel of the material'—these are controlled by a physical and psychological experience of practical work such as only daily contact with building on the site can bring about. This type of theoretical training leads to formalism which Nervi decried.

It is because Nervi mixed the two roles of engineer and contractor, that he achieved a free mastery and control of the structure (Fig. 3.19). Today this knowledge is generally split between different individuals, who may work together but are in constant danger of becoming isolated specialists, and Nervi was the last man to refuse to cooperate in a team. Such was the case when in the UNESCO project in Paris he cooperated with Marcel Breuer and Bernand Zehrfuss. It did not prevent him from giving his utmost, from all his energy to the common task; the three men were so complementary to each other that the danger of a lifeless compromise was overcome. The same applied to the Pirelli skyscraper in Milan, where he collaborated with Gio Ponti.

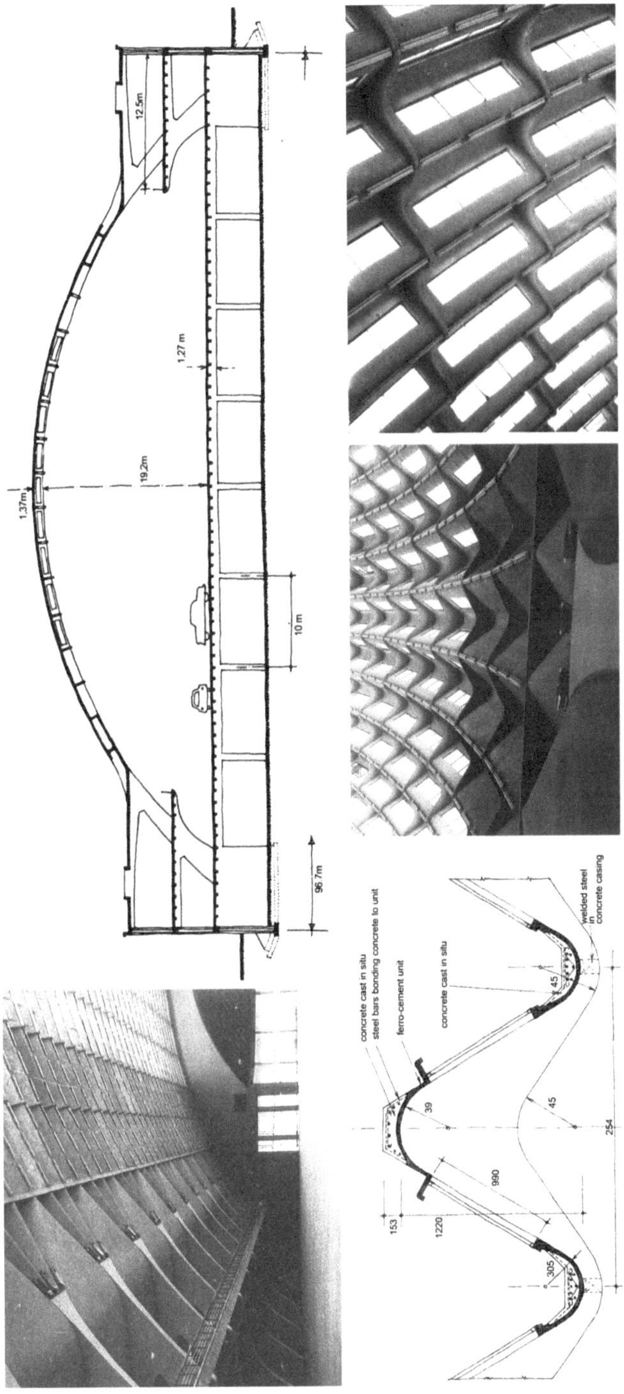

Fig. 3.19 Turin Exhibition Hall B (1947–1954). A prime example of Nervi's use of prefabricated units of ferro-cement and the associated technology over large spans. In this case, the lightness of structure airses from following the isostatic lines of the vaulted roof and combines the potential for elegance through the use of ferro-cement. Cases of the use of this material can be found in his other work, confirming the remarkable strength and lightness of this new method of construction, no less than its notable economic advantages and its great adaptability to architectural forms

From his legacy it becomes clear that Nervi could not help to strike the balance, so that he stands serenely above the mass of the less creative, the sceptics and conventional practitioners. The exceptional example of Nervi remains the ideal foundation for aesthetic appreciation where the ordinary individual invariably can only understand a limited aspect of the whole.

At the time when Nervi was at his peak as a pioneer of structural and constructional techniques and landmark buildings, commentators were presenting points of view that were sometimes sceptical of his achievements and guarded about praise. However, other commentators wrote euphorically that Nervi's technical innovation and demonstration of the correlation of aesthetics and technology was due to familiarity and keen insight into the nature of materials, craftsmanship and methods of assembly. In his hands, the ordinary material of concrete became extraordinary. In his buildings, there exists a union of material, form, structure and pattern, and in these terms his buildings approach the organic. Masterful at making the flow of forces visual, Nervi's building are not the inevitable conclusions of formulaic thinking. Naturally a great and beautiful structure is the end in view, [but by] designing in reinforced concrete it is essential for economy to lead the forces down to the ground by direct thrust. (Jones 2006a).

Both Culmann and Wolff understood the relationship between material patterning, function and performance. Similarly, the work and thinking of Pier Luigi Nervi, (1891–1979) is rooted in the conceptual, creative and pragmatic understanding of material properties and their visual and structural patterning. Nervi's technical innovation and demonstration of the correlation of aesthetics and technology was due to familiarity and keen insight. Reason, emotion, decision-making, authorship convene in his work. It is this agreement that was causal to Nervi becoming known, during and after the 1930's, as a technical artist."
Clifford (2011)

Nervi was still the archetypal builder for whom the ends and the means fuse into harmonious creation. Through insight and experience, the correlation between aesthetics and technology was dependent on the nature of materials, craftsmanship and methods of assembly.

Nervi was active from 1935 with commissions from the Italian authorities before WWII. In 1940 the Italian authorities commissioned a hangar. At the time the need for economy on materials and timber was even more acute, that is why, on the basis of experience which Nervi had acquired he decided to simplify and lighten the structure by designing the ribs as a lattice which enabled him to make use of prefabrication. The construction proved fairly simple. The joints were made by welding the steel and placing high strength concrete in situ in the space left at the junction of four units. The results were excellent, for the eventual six hangars, as was observed before the Germans destroyed them on their retreat by demolishing the supporting columns, but even after the fall of the roof, the majority of the joints remained intact. On the success of that earlier work he was later commissioned to design and construct Hall B for the Turin Exhibition, and later Exhibition Hall C and the Palazzo del Lavori (Labour Palace), both in Turin.

A most important characteristic of Nervi's works is that they are the result of a constant endeavour to find the most efficient design from a technical and economical point of view. Nervi's conviction that the outward appearance of a good building cannot, and must not, be anything but the visible expression of efficient structural and constructional reality, is strengthened by a critical appraisal of the most significant architectural works of the past as well as the present. In other words, form must be the result, and not the initial basis of structure.

It was the conditions at the time that led Nervi to work on the construction method which he called '*ferro-cemento*'. The method is based on the principle of a very thin, highly reinforced slab obtained by forcing a very good quality mortar of cement and sand through several layers of steel mesh and small diameter bars, joined together to form a section. Mortar is placed either by hand or by vibration and the result promises not only exceptional flexibility and strength, but freedom from cracking of the slabs so obtained. Since the mortar is held by the mesh, the formwork can be simplified, or even done away with altogether.

Earlier constructions gave Nervi the experience to attempt the roof of the Turin Exhibition Hall B (1947–1954) using the advantages that ferro-cement construction methods offered. In this case he was invited along with several others to submit a design and tender for the Exhibition Hall to replace the Palace of Fashion destroyed during the War. The problem was interesting, not only because of the dimensions of the Hall with just over 100 m span, but also the very short time allotted for the execution of the work, which had to be completed within seven months. The climate of Turin complicated the problem further.

Exhibition Hall B was undertaken with Roberto di Ruffia and the contractors were Nervi with Bartoli. It was a fitting example where the vault was achieved using special wave-like elements in prefabricated ferro-cement, the structural innovation which became associated with Nervi's many other structures. Nervi was convinced that his ferro-cement technique would attain the necessary stability if corrugated and would enable the use of precasting, and the manufacture of the roof units while the floors and supporting structures were being built.

On this basis he designed the roof structure with corrugated units 2.54 m wide and 4 m long linked by site-cast reinforced concrete ribs in the summits and hollows of the waves. In order to be as light as possible, the units were of ferro-cement (thickness 19 mm) and they would become monolithic through reinforced concrete ribs cast insitu located at the peaks and troughs of the corrugations. In this way ferro-cement units would act as junction units between the insitu ribs which in turn would take over the main structural work. The units were closed at each end by stiffening diaphragms and the adjacent units were joined together by 38 mm of mortar placed insitu. The corrugated roof was connected to the main supporting columns at 7.5 m centres by fan-shaped ferro-cement units springing from inclined reinforced concrete elements.

During construction the units are supported on scaffolding and require no actual formwork. Moreover, the units may be made in any shape and, provided allowance is made for the formation of the ribs, they lend themselves to the expression of a wide range of architectural form.

For the design of the nearly 40 m half-dome at the end of the Exhibition Hall B in Turin Nervi used a method, based on precast units, which he had studied and actually use in small scale structures immediately after the war. The semi-domed apses received small, lozenge-shaped ferro-cement panels (formelle) without the need for double formwork, as would have been the case with ordinary reinforced concrete.

The casting of the units proceeded without any difficulty and without the need for double formwork, as would have been the case with ordinary reinforced concrete. This method, he claimed , was inspired by the need for economising in timber which was extremely scarce in Italy at that time. The method is suitable for the construction of vaults or domes and consists of filling in the space with units measuring approximately 2 m by 4 m. The units are 19 mm inches thick and cast in concrete moulds which in turn are constructed on a model reproducing a section of the vault or dome to be built. The edges of the units are so shaped that when placed side by side they form channels 10.16 cm wide between them, which are filled with in situ reinforced concrete and form a network of supporting ribs that completes the structural system.

The method of construction with precast corrugated ferro-cement units Nervi had proved to be adaptable to large span domes and he prepared many schemes emulating the use of this technique. The assessment of Nervi's ferro-cement structures illustrates how the problems of structural efficiency, economy, thermal insulation, provision of air-conditioning ducts, natural lighting and, more especially, acoustics and sound absorption (very important factors in large buildings) have all been simultaneously solved. This underscores Nervi's philosophy that a design that is sound structurally is generally satisfactory in every other way.

Inaugurated in 1948 and publicised as "the most beautiful building ever built in Italy", the Turin Exhibition Hall B attracted the attention of specialized international press in early 1949, when it appeared on the cover of *La Technique des Travaux*.

The organising body in charge of the Turin Exhibition Hall B subsequently invited Nervi to submit another design and tender for a new hall, Exhibition Hall C, with a ground plan of 55 m by 165 m. Again, the time allotted for the completion was very short being only 4 months. This was an opportunity for Nervi to develop his precasting system further where it could be carried out conveniently in the base of the adjoining main hall.

The Palazzo del Lavoro (Labour Palace) in Turin was designed and built by Nervi between 1959 and 1961. The exhibition building of 47,000 m^2 was a significant Modern Movement masterpiece, and built for the Centenary of the Italian Unification of Italy celebrations—Italia '61—in order to host the international section dedicated to the myths of labour and technical progress. Being an important example of modern architecture in Italy, it is also of historic interest as a

Fig. 3.20 Kursaal at Ostia designed in 1950. The roof consists of prefabricated units and in situ ribs similar to the half-dome construction at the end of the Exhibition Hall B at Turin and used later at the Palazzetto della Sport in Rome. The smooth underside of the prefabricated parts forms part of the ceiling while the upper side is roughened to improve the bond to the in situ concrete topping

"symbol of an extraordinary typological exactness and of an extreme constructional coherence in terms of absolute integration between structural and architectural invention" (Chiorino 2016).

Nervi used his adaptable method of construction many times for curved structures and always with excellent results. The new method of construction of roofs, based on ferro-cement moulds mounted on a travelling scaffolding, movable both horizontally and vertically, enabled the freeing of the construction of the ribbed roof from the restrictions imposed. This new freedom has made it possible for the design of the roofs to follow the statics lines, and make the most efficient use of materials. The harmonious effect and aesthetically satisfying result of the interplay of ribs placed in this way is a clear reminder of the mysterious affinity to be found between physical laws and our visual senses.

The versatility of Nervi's ferro-cement construction method was demonstrated in a project of entirely different character to the Turin Exhibition Halls. The Kursaal at Ostia built in 1950 is a building with a lyrical aesthetic designed for recreational use (Fig. 3.20). With its single column central to the circular plan, the curved soffite becomes a parasol that shelters the interior, and frees the perimeter for uncluttered views of the ocean beyond.

Prior to the deployment of the ferro-cement construction method for buildings, at an earlier stage it was applied for the quick and simple construction of small ships of tonnage not exceeding 500 tons. In 1943 work had started on three motor transport boats for the Italian Navy and one sailing ship, with auxiliary motor for the private industry. The hull of the yacht 'Irene' later built with this method is 38 mm thick and that of other hulls of less than 25 mm thick.

With Nervi's legacy established, he is a worthy claimant to be amongst the masters of the structural aesthetic. In 1960 Pier Luigi Nervi was awarded the Royal Gold Medal.

In investing Nervi with the Medal, the President said:

> Professor Pier Luigi Nervi, by specific authority of Elizabeth II, Queen of this Realm and of the Commonwealth, I place upon you the Royal Gold Medal, which is a symbol to the world of our appreciation to you for your great contribution to architecture. (RIBA 1955).

On receiving the Award, Nervi spoke in French. "Mr President, ladies and gentlemen, my dear colleagues, I am, as you will readily understand, greatly moved by this ceremony, which has touched me deeply. It is difficult for me to find words in which to thank Her Majesty the Queen, you, Mr President, and the Royal Institute of British Architects for the high and noble distinction that I have been given. I believe that beyond the honour done to me personally you wish to show your appreciation of the extraordinary possibilities of the material I have studied all my life. Reinforced concrete is indeed a method of construction which has enormous possibilities, which we are only now beginning to understand after using it for more than 40 years. We are still finding new possibilities in it from the technical, from the architectural and from the structural points of view, and I am certain that in the future it is this system of construction which will be found to be the most rich in its structural and above all in its architectural possibilities.

Even in my native language it would be difficult for me to express adequately my feelings on this occasion, and in French I find it even more difficult, while in English it would be absolutely impossible. I ask you therefore, to appreciate for yourselves the feelings I find it impossible to put into words" (Nervi: RIBA Journal May 1955).

The President, in closing the meeting, called attention to the exhibition of the work of Professor Nervi which was being held in the Institute.

3.7 Oscar Niemeyer (1907–2012): National Congress Centre, Brasilia, Argentine

Oscar Niemeyer, the Brazilian architect of the Modernist period, has been represented as a free spirit through his inventive approach towards the built form. Widely acclaimed internationally his major work was the government buildings in the new city of Brasília that was planned by Lúcio Costa in 1956 to move the capital from Rio de Janeiro to a more central location.

Niemeyer's signature style is associated with the use of curved surfaces, often called 'free forms'. Where other Modernists postulated the importance of elements of control, Niemeyer preferred to leave the philosophical background to his designs to others, giving the impression that every individual building had been the result of a spontaneous act or inspiration. The sensual quality of some of his work can be ascribed to his frank admiration of natural forms: "I am attracted by free-flowing, sensuous curves. The curves I find on the mountains of my native land in the sinuous course of its rivers, in the clouds in the sky, and in the body of the beloved woman" (Petit 1995). Niemeyer's approach was considered by some to be the

Fig. 3.21 National Congress Building in Brasilia, Brazil (1964). The architecture of Oscar Niemeyer became the expression of a new era of democracy in Brazil in 1945. Commissioned to plan a new capital city in the interior of the country, Niemeyer expressed progress and modernity in his designs. The Congress building is located at the head of the vast wing-shaped city plan for Brasilia. The recurrence and preponderance of a proportional system in Niemeyer‘s architecture shows a correlation between his individual design language and a mathematical structure and shows the Golden rectangle as a generative module for his architectural compositions

emergence of a grammar during which functional aspects were not taken into account, instead schematic representations of abstract shapes were aimed at the correlation between the regulating lines of the forms enclosing the volumes, and a depiction of the operations necessary to determine the structure.

Niemeyer's friend and patron, Juscelino Kubitschek, was elected president in 1956 when the concept of a purpose-built capital city in the interior of the country was advanced in earnest. The master plan for Brasilia was opened to national competition, and Niemeyer was a member of the jury. The city planning was awarded to Niemeyer's mentor, Lúcio Costa, in 1957.

The National Congress Building, being located at the head of the vast wing-shaped city plan for Brasilia designed by Costa, and designed in the modern Brazilan style, is given pride of place (Fig. 3.21). The Brazilian Congress is bicameral, composed of the Federal Senate (the upper house) and the Chamber of Deputies (the lower house). In terms of context, being at the head of the Monumental Axis, Niemeyer's design for the National Congress building is fittingly responsive to its important stature amongst the other government buildings.

The architectural forms comprising the building are strikingly articulated to provide the monumental scale necessitated by the expansive vista and are not only determined by function. Rising above the flat roof, the cupola over the Senate chamber taking the form of a shallow parabolic dome, and the bowl-shaped inverted dome over the Chamber of Deputies demarcates the assembly chambers of Brazil's bicameral legislature. The ramp from the elevated driveway is appropriately centred on the majestic twin tower blocks.

At the inauguration of Brasília, President Kubitschek proclaimed the city a national utopia. Other supporters were similarly overzealous. In his book, *Brasília*, published in 1965, Stäubli (1965) concluded: "The principles, originality and perfection of the designs of Lúcio Costa and those of Oscar Niemeyer are so unprecedented and striking that they must be evaluated as a substantial advance in the sphere of modern architecture which established the position of their creators among the great men of modern architecture."

Later studies set out to determine the recurring strategies of proportion in the geometric ordering in Niemeyer´s designs that were structured and characterised by his architectural language. From a twofold analysis of 30 buildings designed by Niemeyer in different periods of his career (1943–2003) it was observed that: "In its first part, the Golden rectangle is confirmed as a generative module for the sample; in the second part, a Shape Grammar model is used as to identify the generative code of five buildings of the sample. This generative code may constitute the basis for the computational formulation of synthetic grammars whereby architects' languages could be developed, taught and automatically reproduced" (Mayer et al. 2007). Mayer observes that as a precursor to the contemporary advantage of digital computation, "It could then be argued that any procedure searching to identify a subjacent proportion system, to a certain extent reproduces the geometric schema used in the design process whereby the architect, in regulating spatial relations of the building elements assigns harmony and unity to the architectural composition". As an approach toward understanding the language and design of complex shapes, although valid for the Niemeyer's work at the time, this analysis predates the algorithmic and mathematically based design methodologies available to more contemporary architects.

In considering the underlying concepts of freedom and limits existing in Niemeyer's architecture, finding that "different control mechanisms, such as the proportional system based on the Golden section used in the conceptual stage of his designs, demonstrates that he applied concepts that became more sophisticated as his career progressed" (Mayer et al. 2007). The recurrent proportionality between the parts of the building suggests that underlying rules of proportion in Niemeyer's architectural grammar can be demonstrated.

To further unravel the complexity of Niemeyer's architectural forms, a team comprising Mayer et al. (2007) determined that using only the Golden section rectangles to "decompose the buildings in modules, abstract compositions of Golden section rectangles in different scales…could be considered as the Shape Grammar model related to the proportional system used in Niemeyer's designs". Without the facility of contemporary computation technology, an ingenious method of analysis was programmed to test whether the Golden section rectangle worked as a topographical reference for Niemeyer's design vocabulary. From analysing the composition of rectangles to generate the final shaping, shape rules were developed based on the composition of rectangles that generated the final shape through the transformation of the initial shape. These derivations were represented by capital letters (in this case T,R,I,S,E) in a generation code identified as 'Shape rules' which were used in 'strings' to represent similarities and differences in generative processes and hence in the final shape aspect of Niemeyer's design grammar in five buildings. Spatial labels were used to indicate points in the grid where the shape rules were applied (initial and final stage) and as the insertion point of rectangles. To develop the grammar further by the unfolding variants of the rules required changes in the sequence of rules, where some adjustment had to be made in the grid to derive the design schema.

Although evidence exists through manual analysis of the recurrence and preponderance of a proportional system and correlation of Niemeyer's architectural language with a mathematical basis, the complexity in increasing the rules and thereby refining the grammar requires computational verification.

Niemeyer's design language thus considered in terms of a proportional system, relied on the expertise of yet another discipline, namely that provided by structural engineer, Joaquim Maria Moreira Cardoso, who was the engineer responsible for the calculations that allowed the construction of the most important monuments of Brasilia, designed by Niemeyer who ranked Cardoso as "the most Brazilian cult that existed".

In *Solid States: Concrete in Transition*, editors Bell and Buckley (2010) allude to the symbiotic relationship between architect and engineer—Niemeyer and Cardozo—who had jointly executed the 'palaces and temples of Brasilia', describing the concepts as: domes resting effortlessly over a floating platform; spectacular exoskeletons sheltering the theatre and the cathedral; and "Niemeyer and Cardozo having pursued an ethereal, anti-monumental monumentality by capitilising on the hybrid materiality of reinforced concrete." They also recall that 'rationalists' denounced these forms as "concrete-clad steel, but in fact that they were concerned with beauty and wonder."

> "Perhaps, because we are taking part in it ourselves, we do not sufficiently realise how profound is the change between architectural concept of the first part of the twentieth century and of the present day, and how important it is to have freed architecture from a complex of rules and traditions which, however much they correspond to structural reality at the time of their inception, have gradually lost any significance and become sterile formalism."
> Nervi (1955a)

Cardozo was a polymath, known as a poet, short story writer, civil engineer, designer, publisher of magazines specialising in art and architecture, as well as being a university professor. He began studying engineering in 1915, graduating 15 years later. The legacy of his brilliance as a structural engineer includes the construction of the whole of Pampulha and palaces of Brasilia: the National

Fig. 3.22 University of Constantine, Algeria (1969–1972). Of note is Niemeyer's understanding of the importance of site context. At the Constantine University, the educational complex buildings are arranged around a vast public space and the more than forty buildings that formed the original project, were incorporated into two blocks. In this way, the freeing of the campus plaza from buildings allowed the simplicity of architectural forms to be offset to the best advantage. Prominent amongst the structures that occupy the campus plain is the lecture hall, a magnificent auditorium with a its reinforced concrete shell-like roofs spanning 24 m, each supported by and sharing a beam on one side. Seen from the air the structural aesthetic is evocative of the wings of a bird

Congress building, Brasilia Cathedral, the Presidential Palace, the Alvorada Palace, and the NS Church of Fatima. He was exonerated after a controversy at the time of the collapse of the fast-tracked project, the Pavilion Gameleira in Belo Horizonte, as it was later proved there were no calculation or structural design errors by the structural engineer.

A specialist in the calculation of structures, Cardozo masterminded the realisation of the unusual, complex shapes designed by Niemeyer. He did not try to adapt them to what was established construction norms for buildings, instead he looked at what was possible through mathematics and the materials available for it. "The two shared the idea that architecture was, first of all, artistic, and it was possible to reconcile it with poetry and engineering. The engineer realized the unusual shapes of the architect and did not try to adapt them to what was established for buildings, looking at a project in order to make it possible, with mathematics and materials available for it" (PET 2013). Cardozo and was the master behind the complex and geometric forms at the educational campus of the University of Constantine, commissioned by President Houari Boumedienne (1965–1978) and now called University Mentouri Constantine (Fig. 3.22).

Situated on a plateau with panoramic views of the city of Constantine, the third largest city and about 640 km east of the Algerian capital, Algier. As in the University of Brasilia, Niemeyer combines in a single volume all the faculties with the complex of educational buildings arranged around a vast public space. The piece de resistance is the "magnificent auditorium with a span of 80 ft, supported by a beam located on the reinforced concrete dome. . .that evokes the wings of a bird.

In this project, Niemeyer's intuitive aesthetic was again demonstrated at the auditorium through the technical possibilities of using shell concrete over large structural spans. For economic reasons and for safety against earthquakes, the buildings were constructed using prefabricated and precast structural parts, which also allow large spans and create large overhangs" (PET et al. 2012).

The design for the auditorium of the University of Constantine in Algeria was 'pure bravado', for which Joaquim Cardozo should also receive recognition as the structural engineer (Bell and Buckley 2010). Cardozo, provided the expertise through which the originality of Niemeyer's conceptions could be realised in real terms, and his virtuoso structural abilities were acknowledged at the time by Niemeyer. Nervi paid tribute to the structural virtuosity of Niemeyer's compatriot, Joaquím Cardozo, as being in "the vanguard of Brazilian engineering," allowing him the "artistic freedom of his architecture".

Niemeyer was recipient of the Pritzker Architecture Prize in 1988, an award modelled on the Nobel Prize, with the final selection being made by an international jury of awardees with all deliberations and voting taking place in secret. Hundreds of nominees are considered year to year from countries all around the world.

Together with Oscar Niemeyer the architect, Cardozo the engineer was a 'master of the structural aesthetic'.

3.8 Jörn Utzon (1918–2008): Sydney Opera House, Australia

The Sydney Opera House's signature sail-like shells evoke Jörn Utzon's original conceptual sketches with a powerful sense of *genius loci*. Viewed from every angle, including from the air, there is a sense of excitement, the shimmering white shells presenting the image of a boat under gigantic sails. Often compared to other romantic and exotic icons of architecture such as the Taj Mahal, the Opera House was Utzon's recognition that the siting offered a unique opportunity requiring a sculptural statement, rather than a conventional design.

Utzon studied the topographical maritime charts and photographs of Bennelong Point, realising that it could be seen from many vantage points. He approached the roof design as a 'fifth facade', thereby ensuring that the building would be the striking finite solution that it is, and he described the Sydney Opera House as having the same attributes as large cathedrals…, "where the structure is also the architecture. The same is seen in Chinese and Islamic architecture, although with different expressions" (Utzon 2002). His original submission "was little more than some rough sketches and was rejected out of hand by the selection committee. World renowned architect Eero Saarinen was asked to be the final judge of the design competition. He rescued Utzon's discarded design from the trash and declared it the winner" (Herron 2015). However, his expressive and iconic architectural statement

Fig. 3.23 Sydney Opera House (1959–1973). Referred to as Utzon's paradigm, Nature was the source inspired the design of the Sydney Opera House, one of the most original and technically innovative, iconic buildings of the twentieth century. Few buildings match the awesome visual 'rightness' of the design, viewed from any angle and the result of Utzon's vision in recognising that the site at Bennelong Point needed to be understood as a unique context. The building stands as a gigantic sculpture with its spherical geometry and consistent curvature throughout, making the cluster of curved white roofs in his competition scheme undoubtedly the most striking feature of the design. Utzon wanted the shells to be constructed in a thin concrete membrane structure, this technology being very popular at the time. The challenge was to find the appropriate geometry for the paraboloid egg-forms of the conception

in the original sketch did not satisfy all commentators—"a magnificent doodle" wrote Australian art critic, Robert Hughes (Murray 2004).

Mikami (2001) observed that "on the competition entry drawing, the shapes of the roof vault had no geometrical definition at all. It was a fanciful and sculptural form drawn by free-hand. Many people felt the shapes of the roofs reminded them of fully blown sails of large sailing boats. The impression it gave was of a lightweight, almost semi-transparent curved membrane floating in the air", but initially Utzon had not been able to arrive at a satisfactory solution to the physical realisation of the paraboloid egg-like roof structures (Fig. 3.23).

In the realisation of the project, how to bring Utzon's original flight of imagination into physical reality, became the subject that prolonged tensions between all roleplayers. According to Utzon when he penned the design for the Sydney Opera House in 1948, sustainability was one of his primary objectives. "Sustainability is part of the Sydney Opera House's DNA," agrees Louis Herron, chief executive of the Sydney Opera House. "Architect Jørn Utzon was inspired by Nature when designing the Opera House—by the harbour, by organic forms, colours and light—

and sustainability is inherent in his design philosophy. He believed that architecture should grow out of the landscape and resemble a single organism" (Herron 2015).

At the time of the review of the initial concept, Pier Luigi Nervi (1891–1979) pointed out that Utzon's proposal, having been favoured for such a complex project, should have more specific information in the architectural terms. Engineering data on its structural and constructional feasibility should have been included, and this lack of structural design at the early stages of the project was indeed the primary cause of the major cost and eventual schedule overruns. In fact the views expressed by Nervi were virtually indistinguishable from that of Ove Arup, who became the structural engineer on the project. Like Arup, Nervi was opposed to considering the structure solely on the basis of its external appearance, and he held that architects of the future "Must possess—and synthesise—aesthetic sensibility, profound understanding of structural needs, and a precise knowledge of methods, possibilities and limitations of constructural techniques" (Jones 2006b).

The greatest almost insurmountable challenge was how to apply a structural geometry to the organic forms and the construction of the shells themselves. The roof structure was "of such magnitude and complexity that cannot be built without explicit geometry that can be expressed mathematically. Without such a mathematical model, it is not possible to calculate loads, stresses, and rotational forces that the vaults would be subjected to, and to estimate the impact of wind and temperature changes in their stability (Hahn 2012c). Arup instigated a solution that Utzon initially endorsed: that the roof structures could only be built as a sequence of curving ribs from narrow at the base and widening as they rose. However, the geometry remained a problem.

Utzon's flash of inspiration came when he realised that his shell vaults were the equivalent of curving triangles drawn from the same sphere. All the triangles for the design of Utzon's shells are taken from the sphere of radius of 75 m. This geometry became the basis for the final design five years after the official announcement that Utzon had won the competition. Fjeld (2009) evoked the poetic cultural dimension of this decision, by suggesting that "When Utzon was to realise his Sydney Opera House, he went to the geometry of the ball. By cutting sections from a ball he found the exact shape for his shells. It was as if he moved into the past, destroyed the dome of the cathedral, and by gathering the pieces left on the ground, he suddenly had the tools to realise the poetic dream of the present."

Utzon's concept owed its oeuvre to the earlier work of Felix Candela, Robert Maillart and indeed Pier Luigi Nervi where parabolic curves spanned large spaces. The thin concrete membrane of the TWA Terminal at the airport, New York by Eero Saarinen had demonstrated what was possible. Intense investigations in collaboration with Ove Arup and his engineers, including advanced model simulations at Southampton University, preceded a breakthrough decision to adopt a spherical geometry, the most effective way in Nature to contain the greatest volume with the least surface area.

As the first researcher to have full access to the vast private Arup archives, Jones tells the complete story of Ove Arup's extraordinary life, his social, aesthetic, and environmental concerns, and his practical contributions during a period of political and technological upheaval. Jones also offers a revelatory new account of the Sydney Opera House and the character of those involved in its complex construction.
Jones (2006b)

The spherical geometry of the varying sized roofs with the same curvature, Utzon attributed to the segments of an orange, while the structure of palm fronds provided models for resolving the complex construction of the fan-like ribs that support the roofs. (Carter et al. 2013a) (Fig. 3.24). As work on the design of the shells progressed a common denominator was established, namely, exploiting the nature of spherical geometry could be applied to the curvature throughout. A large amount of scaffolding and shuttering for the interior and exterior forming of the concrete shells could be dispensed with allowing subdivision into ribs, and again into smaller elements that could be precast in formwork representing the largest component and assembled in pieces into position (Carter et al. 2013a).

Within a year of working on the roof design, Ove shared long conversations about the design and modelling of shells with various engineers of note during their visits to London, and with Pier Luigi Nervi whom he already knew and greatly admired: all the overseas engineers he consulted had severely criticised Utzon's published sketches. As a leading engineer of the twentieth century and founder of a firm of consulting engineers that brought to practical fruition such iconic structures as the Sydney Opera House, Ove Arup (1895–1988) was devoted to a 'common-sense' reform of engineering and architectural practice and pioneered the way for modern architects and engineers to collaborate successfully and responsibly on large projects. In some circles he, as did Nervi, added controversy to the debate on the intrinsic merits of good architecture.

I have also observed that the contribution of technical-constructional knowledge is only effective if it is brought into collaboratoin with the architect from the inception of the scheme; as with living creatures, it is very difficult to eliminate in the initial stages any initial deficiency or malformation. It is clear that the whole field of construction, which stretches from the cultural formation of the architect on the one hand to the industrial organisation of construction on the other, is in a state of rapid and progressive development which demands a bringing up to date of both ideas and techniques. All efforts of those who have at heart the progress of architecture, will still be inadequate to define, to study and to solve so many and such complex problems.
Nervi (1955b).

Fig. 3.24 Sydney Opera House sail-like shell domes on the massive podium that almost covers the site. The shells were designed to face in all directions, and thus the spheres required a given radius of curvature in order to provide for the assembly of the curving triangles. This became Utzon's solution. The large window areas that resulted bring both light and vistas of Sydney harbour into the lobbies of the complex. The podium extends over an area of 116 m by 186 m and accommodates the Opera House's chamber music hall and experimental theatre as well as other facilities and services

To solve the structural challenges of Jörn Utzon's Sydney Opera House, Arup explored the best way to deal structurally what he called "Utzon's fanciful and sculptural form drawn by free-hand". Yuzo Mikami, the Japanese architect who first worked with Utzon and then subsequently with Ove Arup on the design of the Opera House, writes in *Utzon's Sphere*, his own account of the project: "Many people felt the shapes of the roofs reminded them of fully blown sails of large sailing boats. . .The cluster of curved white roofs in the competition scheme were undoubtedly the most striking feature of Utzon's design. He wanted them to be constructed in a thin concrete membrane structure, this technology being very popular at the time" (Mikami 2001).

Carter et al. (2013a) observed that the work of others, such as Pier Luigi Nervi who "played a role in opening Utzon's imagination to the new technical and aesthetic possibilities of large parabolic reinforced concrete roof spans." Descriptions of the Sydney Opera House lyricise its metaphoric imagery, its tectonic integrity and and its intentional origins through Utzon's belief in Nature-inspired organic architecture. Post-modern architecture that followed the era of the Opera House was also represented in high-tech architecture and by contrast has been defined as two-dimensional, and lacking sculptural merit.

In 2003, Utzon was awarded the Pritzker Architecture Prize, architecture's highest honour. The award citation states: "There is no doubt that the Sydney Opera House is his masterpiece. It is one of the great iconic buildings of the twentieth century, an image of great beauty that has become known throughout the world—a symbol for not only a city, but a whole country and continent."

3.9 Frei Otto (1925–2015): Munich Olympic Stadium 1972

A review of innovative structural design down the centuries should emcompass the work of all who dispensed with conventional ways and embraced the "modern emphasis on honest structural expression [that] has resulted in more sincere and innovative interpretations of Nature in spatial structures. The exemplary Nature-inspired tectonic architectures of Jørn Utzon, together with comparisons to the works of amongst others; Gaudi, Frei Otto, Nervi. . ." (Carter et al. 2013b) are representative of the individuality that has emerged from a range of paradigms.

Frei Otto, born in Siegmar, Saxony in 1925, spent his long career researching, experimenting, and developing a most sensitive architecture that has influenced countless others throughout the world. As a boy he had built many aircraft models and later pursued his passion by becoming a Luftwaffe pilot. He was later shot down and incarcerated in a French PoW camp near the end of WWII.

Otto made his mark with a number of impressive ultra-modern and super-light tent-like structures using new materials, beginning with the West German pavilion, designed with Rolf Gutbrod, for the 1967 Montreal Expo. His tensile structures draw on natural phenomena—from birds' skulls to soap bubbles and spiders' webs, or could be likened to the roof of a tent, where fabric hangs between two points in tension and adopts a catenary shape, versus a cabin, where a strict geometry determines the final form. In using highly innovative lightweight materials, Otto's genius for tensile structures was brought to the attention of the world.

In his quest "to discover the lightest, strongest, most responsive and elegant structural solutions for buildings [Otto] drew his imagination from across history and through Nature. In his publication *Ancient Architects* in 1994 [he] searched for the most inventive use of materials by the very earliest architects", followed by his "brilliant and much-reprinted monograph *Finding Forms: Towards an Architecture of the Minimal*" (Glancey 2004).

Weblike structures, such as the conventional tent, have a high load-bearing capacity yet the structure itself is lightweight. Spider silk has been shown to be stronger and more elastic than Kevlar, for some time the strongest man-made fibre. Otto pioneered the return of the traditional tent for lightweight, sometimes temporary, adaptable web-like buildings that have emanated from the lessons of Nature. He embraced the definition of architect to include researcher, inventor, form-finder, engineer, builder, teacher, collaborator, environmentalist, humanist, and creator of memorable buildings and spaces. The extraordinary depth and breadth of Otto's

Fig. 3.25 Munich Olympic Stadium (1972). The sweeping canopies of acrylic panels floating over the stadium, were stabilised by steel cables and were an innovative system used for the first time on a large scale. The design with its peaks, echoed the Alps and symbolised the new democratic and optimistic Germany. Views out towards the surrounding landscape added to the sense of place. Otto's low-cost, and sometimes even temporary structures were a direct reaction to the presumptuous idea that any building is forever, or that architecture was a tool for promoting an ideology

research work in the field of lightweight structures that are adaptable and changeable, embodied the careful use of limited resources that equipped him to become one of the greatest architect-engineers of all time (Fig. 3.25).

On the aesthetics of fabric tension structures, Khan (1980b) observes that "…structural simplicity and elegance have always been the essence of the overall architectural expression of these buildings. The pioneer work done for the Munich stadium for the Olympics of 1972 (by Otto) is an excellent example of this kind of structure."

Designed in collaboration with Gunther Behnisch, the roofing for the Munich Stadium was considered a pioneer in lightweight tensile and membrane construction. Using lightweight, inexpensive plastics or plexiglass strung between complex hardware frameworks to create huge, light-filled volumes that could be easily assembled and disassembled was a departure from the 1936 authoritarian Olympic structures of the Nazi era. The lightweight, open to nature, democratic symbolism of the structural aesthetic was reflected in the official motto: "The cheerful Games".

Another example of the aesthetic potential of lightweight structures, is the Haj Terminal in Jeddah by SOM architects (US) that covers 47 ha and uses the tent form. In the Terminal the tent units are hung from piers at 45 m and soar 20 m above ground reaching a height of 35 m thereby demonstrating the extraordinary properties of teflon-coated fibreglass 1 mm thick fabric interacting with the radiating cables.

Through his specialised knowledge and experience with lightweight tensile and membrane structures, Otto pioneered advances in structural mathematics and civil engineering. He was the founder of the Institute for Lightweight Structures at the University of Stuttgart in 1964 and headed the Institute until his retirement as university professor.

Using Nature as a model, Otto in collaboration with biologist J G Helmke, started studying Radiolaria micro-organisms—within the field of biomimesis—looking for a structural design medium that could account for the state of natural equilibrium for lightweight structures. From this emerged his theory, so-called Form-Finding, which led to his "analogic models, like hanging chains or nets of cables" (Mentegazzi 2014). Frei Otto's research on constructing light-weight, tensile structures through form-finding physical models that performed analog 'material computation' is regarded as a precursor to Parametricism, being the form-finding technique which emerged in the 1990s that has fed into an avant-garde movement in architecture. Otto used "physical processes as simulations and design engines to 'find' form rather than to draw conventional or invented forms…The inherent lawfulness of the engaged physical processes produced a combination of complexity, rigour and elegance that was otherwise unattainable. The power and beauty of this approach was striking." (Schumacher 2016a).

After WWII, Otto studied at the Technische Universität, Berlin, and, as a postgraduate student researched sociology and urban form at the University of Virginia. During that time he visited the leading-edge US studios of Erich Mendelsohn, Mies van der Rohe (both German emigres) Eero Saarinen, Richard Neutra, Charles and Ray Eames and Frank Lloyd Wright, but tensile structures remained his motivating passion. Working with a leading tent-maker, his numerous pioneering structures were given extensive coverage in the world media. (Glancey 2004).

In his citation for the RIBA Royal Gold Medal in 2005 to Frei Paul Otto, the president, George Ferguson stated that the medal is awarded to a person whose influence on architecture has had a truly international effect. On a professional note he added: "He has always been one of my engineering and architectural heroes and has a genuine claim to be one of the real greats of the twentieth century. I hope he will recognise this as the crowning of a truly outstanding and pioneering career."

As a progressive designer of the 1960s and 1970s, Frei Otto's ideas were featured in the Whole Earth Catalog of the time with examples of his work. In 2006, Frei Otto was awarded the 18th annual Praemium Imperiale prize for architecture, by the Japan Art Association, and later in 2015, the Pritzker Architecture Prize, with accolades from the individual juror laureates.

The Pritzker Prize for Architecture Jurors of 2015, themselves laureates, individually recognised Frei Otto's as having inspired modern architecture from which all could learn.

Chair of the Jury, Peter Palumbo cited: "Frei stands for Freedom, as free and as liberating as a bird in flight, swooping and soaring in elegant and joyful arcs, unrestrained by the dogma of the past, and as compelling in its economy of line and in the improbability of its engineering as it is possible to imagine, giving the marriage of form and function the invisibility of the air we breathe, and the beauty we see in Nature."

Richard Rogers: "His work has inspired and influenced modern architecture, as we all learn to do more with less, and to trade monumental structures for economy, light and air."

Frank Gehry: "Frei Otto forever changed the way we think about structure and building. Through his experiments in form-finding, Otto simultaneously affirmed and questioned the conventions of engineering as we knew it, and in the process showed us unprecedented solutions to age old problems—where others saw mass as the solution, he offered lightness. Like the ancients and others that came before him, he questioned the origins of our assumptions by going back to nature and figuring it out for himself."

Zaha Hadid: "The fluidity of Frei Otto's work is as uplifting as it was profoundly inventive—a persuasive manifesto of nature's logic and unity, demonstrating how architectural design and engineering can emulate nature's morphogenesis. The more our own design research evolves, the more we learn to appreciate his pioneering works. He will continue to influence architects and engineers for generations to come."

Otto's response to being awarded the Pritzker Architecture Prize was telling about his humanitarian values, and about his drive: "to design new types of buildings to help poor people especially following natural disasters and catastrophes...You have here a happy man." After Frei Otto's death just weeks short of his 90th birthday, tributes poured in from the world-wide architectural fraternity.

3.10 Zaha Hadid (1950–2016): Heydar Aliyev Centre, Baku, Azerbaiyan

Zaha Hadid is without a doubt the world's most famous woman in a star-studded line-up of masters of the structural aesthetic, a stratosphere dominated by her masculine peers. It is fitting that, before her death, her life in architecture should be rewarded for the RIBA Royal Gold Medal for 2016.

It has to be acknowledged that the role of the architect in modern times has become modified through the separation of the functions needed to complete the process of building design and construction. Whereas until the advent of the digital age, the architect's brief was not only to conceptualise the design of a building and to produce sets of technical drawings and specifications manually that rendered to the client's needs, it went further—to instruct the builder in the matter of the construction itself. Through contemporary cultural factors within society and the construction industry, the architect has tended to become a specialised member of a team rather than the chief agent, although the brief now embraces wider parameters, such as the significance of the context and issues of sustainability of their buildings. This separation of roles is evident in the modus operandi by which many of the iconic buildings of modern times are conceived and constructed.

Zaha Hadid's singular success as a modern architect of note relied on digital design aids through which her iconic structures could be realised into their fluid material form. Her Utopian designs based on her philosophical ideology are widely indulged and enjoyed by the public due to their experiential value, a response to the current continuous search for originality especially in the developed world from Europe to Asia and the Americas.

With no intention to detract from Hadid's significant place in architectural history, a study of the very roots of her design philosophy provides greater understanding of the seemingly anarchic structural forms she conceived. As a student at the AA school of architecture in London, Hadid was inspired by Kazimir Malevich (1879–1935) from pre-revolution Russia whose geometric abstraction and fragmentation symbolised a movement called Suprematicism that became the touchstone of her own design concepts and indeed her architectural self. Malevich's system of thought liberated and affirmed her architectural philosophy, allowing her to see space differently and made her think about weightlessness in Art (Sudjic 2014). Hadid began to question existing fundamental parameters, and explained that Suprematism had meant the beginning of a new culture.

Zaha Hadid's architectural design concepts represented by the numerous buildings which could be described as 'frozen in motion', should not, however, be considered replications of Malevich's Suprematist art works, or as radical gestures symbolised by his Black Square that was "like an open window through which the revolutionary spirits of radical destruction could enter the space of culture and reduce it to ashes" (Groys 2013) (Fig. 3.26). However, like Malevich, her work should rather been seen as "announcing the death of any cultural nostalgia, of any sentimental attachment to the culture of the past." Malevich (1924 in Groys 2013)

Fig. 3.26 The Heydar Aliyev Centre, Baku, Azerbaijan (2007–2012) The city of Baku, celebrated by the elites as the centre of cosmopolitan life in the Caucasus, experienced a surge of new wealth from oil and gas projects, and a sizable growth in population. It was logical that this should be symbolised through some tangible form. There could be no better means than an architectural work by a celebrated architect, indeed by a woman. In the post-Soviet era, a stunning pace of construction is found in many Asian cities from patronage from powerful political imaginaries that seek to commission art in the service of state power, apart from political prestige. The Heydar Alijev Gallery to a large extent exemplifies this politico-cultural patronage however, it's main focus is the collective celebration of contemporary and traditional Azeri culture

ventured his theory into architecture itself: "The architect by nature is always abstract. But life sets him the task to build his own abstract forms. Situate them in such a way that useful space for life may form itself amidst them."

The curator of the Malevich Tate Museum in the UK observes that Malevich's "Architectons are models without practical purpose—no doors or windows—a dream world pushing forward into an unknown terrain and developing an architectural utopia" (Sudjic 2014).

Hadid was in the lead for a sea change in architectural language and expression (Fig. 3.27). This is a reminder of that politico-cultural climate—the pre-revolutionary phase of Russian avant-garde within which the figure of Kazimi Malevich emerged before—that represented "the strongest possible medicine against any kind of compassion or nostalgia" (Groys 2013). The analogy is the contemporary cultural environment in which Hadid operated and in which she chose to depart from the "persistence of styles such as modernism, minimalism, post-modernism, historicism and deconstructivism, a hard core of continuous innovation in research and building [that] has stabilised around the new heuristics of Parametricism, and is continuing to proliferate the new style in academic and practice domains worldwide" (Schumacher 2016b). A distinction should be made that her work is the

Fig. 3.27 The Heydar Aliyev Centre, Baku, Azerbaijan (2007–2012). Views that illustrate the extraordinary fluidity of the volumetric form achieved through the use of parametric design principles and the execution using the spaceframe. The engineering design for the 3D free-form architecture of the Heydar Alijeve Centre by Hadid was by Werner Sobek whose deployment of

Fig. 3.28 The Heydar Aliyev Centre, Baku, Azerbaijan (2007–2012) Free-formed exterior cladding of glass fibre reinforced concrete

"opposite of high-tech architecture—she was magic with space, there is a dynamism" (Sudjic 2014).

One of the most encapsulating of Hadid's architectural design approach is the Heydar Aliyev Centre in Baku, Azerbaijan, formerly part of the former Soviet Union, with its distinct flavour of the urbanism and architecture that was heavily influenced by the planning of that era. Since its independence in 1991, Azerbaijan has invested heavily in modernising and developing Baku's infrastructure and architecture, departing from its legacy of normative Soviet Modernism. Zaha Hadid Architects was appointed as design architects of the Heydar Aliyev Center following a competition in 2007. The Centre, designed to become the primary building for the nation's cultural programmes, breaks from the rigid and often monumental Soviet architecture and aspires instead to express the sensibilities of Azeri culture and the optimism of a nation that looks to the future.

Hadid's Heydar Aliyev Centre achieves elaborate undulations, bifurcations, fluid folds, and inflections that transform the banal urban space into an architectural landscape that welcomes, embraces, and coaxes visitors into and through different levels of the interior (Fig. 3.28).

Fig. 3.27 (continued) construction logics and technical systems through advanced computer modelling was based on teamwork and achieved the geometry of the continuous, flowing surface. In principle the concrete structure is combined with a space frame system that resulted in the large-scale column-free interior voluminous spaces. The unconventional merger of structural systems enabled the convoluted and inverted continuity of forms that have realised the fluid geometry of the structural aesthetic, both internally and externally

The structure consists principally of two collaborating systems: a concrete structure combined with a space frame system. In order to achieve large-scale column-free spaces that allow the visitor to experience the fluidity of the interior, vertical structural elements are integrated into the envelope and curtain wall system. Through the use of the space frame, a free-form structure was enabled saving time throughout the construction process, while the substructure was developed to incorporate a flexible relationship between the rigid grid of the space frame and the seams of the free-formed exterior cladding of glass fibre reinforced concrete, as well as glass fibre reinforced polyester. These materials met the demand for the plasticity of the building's design while responding to the functional enclosure of the envelope.

As the most contemporary master of a particular structural aesthetic, Hadid does share the acclaim with a wider team, amongst whom has been team architect Patrik Schumacher who successfully applied parametric method to her designs that he called 'parametricism' (2016a). This method "engaged physical processes [and] produced a combination of complexity, rigour and elegance that was otherwise unattainable. The power and beauty of this approach was striking" (Schumacher 2016b).

Despite endorsement from august juries, Hadid's buildings sometimes faced criticism from commentators in a world where the role of the general public as a valid stakeholder is becoming paramount. Her Heydar Alijev Centre in Baku in the Azerbaijani capital did face a backlash from human rights groups who cited residents forcefully evicted to clear the site for Hadid's cultural centre, and journalists attributing negative socio-political issues to her role as architect. However, in regard to Hadid's choice of architectural expression, there can be no detracting from the impact of her extraordinary designs that push the boundaries of conventionality and technology and building methods. She should be recognised as a force for challenging the perceptions about the limits of building technology and methods—digital and otherwise—and for achieving her design objectives. Hadid received the Pritzker Architecture Prize in 2004. Through her legacy, her place as one of the masters of the structural aesthetic is secure.

At the presentation of the Medal the president of the RIBA cited: "Zaha Hadid is a formidable and globally-influential force in architecture. Highly experimental, rigorous and exacting, her work from buildings to furniture, footwear and cars, is quite rightly revered and desired by brands and people all around the world. I am delighted Zaha will be awarded the Royal Gold Medal in 2016 and can't wait to see what she and her practice will do next" (Duncan 2016).

On receiving the RIBA Gold Medal, Hadid responded to the citation: "I am very proud to be awarded the Royal Gold Medal, in particular, to be the first woman to receive the honour in her own right. I would like to thank Peter Cook, Louisa Hutton and David Chipperfield for the nomination and Jane

(continued)

Duncan and the Honours Committee for their support. We now see more established female architects all the time. That doesn't mean it's easy. Sometimes the challenges are immense. There has been tremendous change over recent years and we will continue this progress. This recognition is an honour for me and my practice, but equally, for all our clients. It is always exciting to collaborate with those who have great civic pride and vision. Part of architecture's job is to make people feel good in the spaces where we live, go to school or where we work—so we must be committed to raising standards. Housing, schools and other vital public buildings have always been based on the concept of minimal existence—that shouldn't be the case today. Architects now have the skills and tools to address these critical issues."
(Hadid 2016)

Other plaudits included those from the handful of designers, thinkers and entrepreneurs who created the International Design Awards (IDA) in 2007 and recognised Hadid's Heydar Centre as a celebration for smart and sustainable multidisciplinary design. Furthermore, the Design Museum of the Year provoked a conversation around the realities of contemporary architecture that the Museum wanted to be a part of adding that: "There is no doubt that Zaha Hadid is a hugely talented architect, and this is what the Museum's Design of the Year award recognises" (Sudjic 2014).

In the design of the Heydar Aliyev Centre, Hadid has exploded the conventional and familiar architectural forms and sculpted a building into organic fluidity that arouses the curiosity, that welcomes and liberates the rigid context from its banality. Other architectural statements that reflect contemporary cultural norms in the developed world, have stretched the boundaries of design through original structural forms, however, Zaha Hadid's work places her in the pantheon of masters of the structural aesthetic.

References

References 3.1

Dethier, J. 1981. *Mud architecture: An old idea, a new future* (pp. 8–33). Based on the exhibition Des Architetures de Terre, initiated and directed by Dethier, J. at Centre Pompidou, Paris, with the Deutsches Architektur Museum, Frankfurt (1981–1982). Thames & Hudson.
Fletcher, B. (1948a). *A history of architecture on the comparative method* (14th ed.pp. 157–158). London: B.T. Batsford.
Hahn, A. J. (2012a). *Mathematical excursions to the world's great buildings* (pp. 31, 32, 36). Princeton: Princeton University Press.
Maguire, J. (2008). *Mali notes and records*. Report on the ArchSoc Tour of November-December 2008.

Online

Bondono, E. (2016). Accessed October 13, 2016, from http://www.jeffbondono.com/TouristInRome/Pantheon.html

El Akkioui, H. (2011). Accessed December 10, 2016, from http://www.sud-maroc.com/2011/02/kasbah-architecture-du-sud-marocain/

Red Fort Complex. (2016). World heritage site. Accessed September 25, 2016, from http://whc.unesco.org/en/list/231/gallery/

Roman concrete. (2016). Accessed October 13, 2016, from https://en.wikipedia.org/wiki/Roman_concrete

Wayman, E. (2016). Accessed October 14, 2016, from http://www.smithsonianmag.com/history/the-secrets-of-ancient-romes-buildings-234992/#Liv6lPVUQgiuc9DJ.99

References 3.2

Alexander, C., Ishikawa, S., & Silverstein, M. (1977). *A pattern language*. New York: Oxford University Press.

Alexander, C. (1979). *The timeless way of building*. New York: Oxford University Press.

Fitchen, J. (1981). *The construction of Gothic cathedrals: A study of medieval vault erection* (p. 124). Chicago: The University of Chicago Press.

Fletcher, B. 1948b. The Hellenic period (700-146BC) and Gothic architecture. In *A history of architecture on the comparative method. 2: Architectural character* (14th Edn., pp. 72, 75, 340). London: B.T. Batsford.

Ramzy, N. S. (2015). The dual language of geometry in Gothic architecture: The symbolic message of Euclidian geometry versus the visual dialogue of Fractal geometry. *Journal of Medieval Art and Architecture, 3*(2), 138, 157, 168.

Salingaros, N. A. (1999). Architecture, patterns and mathematics. *Nexus Network Journal, 1*(2), 75–85.

Online

Chartres Cathedral. (2016). Accessed October 29, 2016, from https://en.wikipedia.org/wiki/Chartres_Cathedral

Roman building materials. (2016). Accessed November 16, 2016, from http://www.mariamilani.com/ancient_rome/rome_building_materials.htm

References 3.3

Hahn, H. (2012b). *Mathematical excursions to the world's great buildings* (p. 99). Princeton: Princeton University Press.

Khan, F R. (1980a). Structural aesthetics in architecture and its social and technological relevance. In *Zeitschrift: IABSE congress report AIPC=IVBH, Kongressbericht Nov* (p. 136).

Mueller, T. (2014a). Il Duomo. Brunelleschi's Dome. How did a hot-tempered goldsmith with no formal architectural training create the most miraculous edifice of the Renaissance? In *National geographic, February 2014*.
Vasari, G (1511–1574): 2012. *Lives of the most eminent painters sculptors and architects*. In De Vere, G. Du C. (Trans.) (Vol. 1, p. 136). London: Macmillan & The Medici Society.
Zunz, G. J. (1997). Working on the edge – The engineers' dilemma. In R. J. W. Milne (Ed.), *Structural engineering: History and development* (pp. 44–45). New York: CRC Press.

Online

Ricci, M. (2016). Accessed November 14, 2016, from http://news.nationalgeographic.com/news/2014/02/140210-duomo-florence-brunelleschi-cathedral-architecture/
Mueller, T. (2014b). Accessed November 15, 2016, from http://ngm.nationalgeographic.com/2014/02/il-duomo/mueller-text
Brunelleschi's Dome. (2014, February). *Detail image: Digital edition of National Geographic magazine*. http://ngm.nationalgeographic.com/2014/02/il-duomo/cutaway-interactive

References 3.4

Ahuja, A. (2016). *Integration of nature and technology for smart cities. 13.1 Biophilic design* (3rd ed.pp. 331–332). Cham: Springer.
Lorenzi, M. G., & Francaviglia, M. F. (2010). Art and mathematics in Antoni Gaudi's architecture. "La Sagrada Familia". *Journal of Applied Mathematics (Aplimat), 3*(1), 126, 128, 129, 130, 131, 137, 140, 143.
Milne, R. J. W. (1997a). *Structural engineering: history and development* (p. 48). New York: CRC Press. T&G Group.
Zerbst, R. (2013). *Gaudi 1852–1926: Antoni Gaudi I Cornet: A life devoted to architecture, Architecture and Design Big Series* (p. 208). Cologne: Benedikt Taschen Verlag.

Online

Barnes, M. E. (2014). *Ernst Haeckel's Biogenetic law (1866)*. Accessed December 12, 2016, from http://embryo.asu.edu/handle/10776/7825

References 3.5

Kidder-Smith, G. E. (1961). *The new architecture of Europe* (p. 92). London: Pelican Books.
In Volume II. *The 1500s to the future. Architecture and mathematics from antiquity to the future*. Switzerland: Birkhäuser Springer.
Evans, R. (2000). Chapter 8. Forms lost and found again. In R. Evans (Ed.), *The projective cast – Architecture and its three geometries* (pp. 303–304). Cambridge, MA: MIT Press.

Online

Ronchamp Visitor Guide. (2016). *Colline Notre-Dame du Haut, Ronchamp – site officiel mentions légales*. Accessed September 12, 2016, from http://www.collinenotredameduhaut.com/discover/the_builders_of_the_hill.1581.html

Chapel of Notre Dame du Haut in Ronchamp and Braga Stadium. KTStudioKT.net

Trachtenberg, M. & Hyman, I. (2016). Accessed November 17, 2016, from http://www.greatbuildings.com/buildings/Notre_Dame_du_Haut.html

Greer, G. (2016). Accessed November 25, 2016, form http://www.theguardian.com/artanddesign/2007/oct/18/architecture3

Chapel of Notre Dame du Haut in Ronchamp and Braga Stadium. (2016). Accessed November 27, 2016, from http://www.greatbuildings.com/buildings/Notre_Dame_du_Haut.html

References 3.6

Clifford, D. (2011). *Project Nervi: Aesthetics and technology* (pp. 74–82). ARCC 2011, Considering Research: Reflecting upon current themes in Architecture Research Carnegie Mellon University.

Jones, P. (2006a). *Ove Arup: Master builder of the 20th century* (p. 196). New Haven, CT: Yale University Press.

Charleson, A. (2006). *Structure as architecture* (p. 22). New York: Routledge.

Chiorino, C. (2016). *Pier Luigi Nervi project. Preserving the legacy of – Enhancing the knowledge about Pier Luigi Nervi*. Chiorino, C. member: Docomomo Italy & advisor: The preservation of the architectural heritage for the Pier Luigi Nervi Project Assoc.. Chair Marco Nervi.

Melaragno, M. (1991). *An introduction to shell structures: The art and science of vaulting* (pp. 175–184). New York: Van Nostrand Reinhold.

RIBA Gold Medal Award. (1955). *Nervi's acceptance speech*. 232. RIBA Journal, May 1960.

Institution of Structural Engineers and Joint Committee on Structural Concrete on 15 October 1955.

Rogers, E. N. (1957). *The works of Pier Luigi Nervi*. vi, ix, x, xi. Preface: Pier Luigi Nervi; Introduction: Rogers E. N.; Notes to illustrations: Joedicke, J. Ttranslation: Priefert, E. New York: Praeger.

Online

Chiorino, C. (2011). *Enhancing the knowledge about Pier Luigi Nervi*. Accessed December 3, 2016, from http://pierluiginervi.org/pier-luigi-nervi-architecture-as-challenge/the-12-architectural-icons-presented-in-the-travelling-exhibition

References 3.7

Bell, M., & Buckley, C. (2010). *Solid states: Concrete in transition* (p. 69). Princeton, NJ: Princeton University Press.

Mayer, R., Martins, G. E. G., & Turkienicz, B. (2007). *Niemeyer proportion rules* (pp. 1, 2, 5). Conference Paper: Delivered at the Fifth International Conference on Mathematical Design. Organiser: Mathematics and Design Association, Blumenau.
Petit, J. (1995). *Niemeyer Poete D'Architecture*. Lugano: Fidia Edizioni D'Arte.
Nervi, P. L. (1955a). Concrete and structural form. *The Architect and Building News, 208*(27), 523–529.
Stäubli, W. (2009). *Brasilia* (p. 10). Koch, A. (1965). New York: Cornell University.
Styliane, P. (2008). *Oscar Niemeyer: Curves of Irreverence* (p. 245). New Haven: Yale University Press.
Turkienicz, B., & Mayer, R. (2015). Oscar Niemeyer curved lines: Few words many sentences. In K. Williams & M. J. Ostwald (Eds.), *Architecture and mathematics from antiquity to the future* (Vol. II, pp. 388–389). Switzerland: Springer.

Online

PET Engenharia Civil UFj. (2013). *Notable engineers, Joaquím Cardozo*. Accessed October 25, 2016, from https://blogdopetcivil.com/2012/12/17-2/engenheiros-notaveis-joaquim-cardozo-2/

References 3.8

Carter, A., Kirkegaard, P H., & Tyrell, R. (2013a). The nature of tectonic architecture and structural design. In da Sousa Cruz, P. J. (Ed.), *Structures and architecture: New concepts, applications and challenges* (pp. 237, 240–243). London: Taylor and Francis Group.
Fjeld, P.-O. (2009). *Sverre Fehn – The pattern of thoughts* (pp. 10–180). New York: The Montacelli Press.
Hahn, A. (2012c). *Mathematical excursionsto the world's greatest buildings* (pp. 239, 261–262). Princeton: Princeton University Press.
Jones, P. (2006b). *Extract from Ove Arup: Master building of the 20th century* (p. 196). New Haven, CT: Yale University Press.
Mikami, Y. (2001). *Utzon's sphere* (pp. 32–33). Tokyo.: Shokokusha.
Murray, P. (2004). *The saga of sydney opera house* (p. 10). New York: Spon Press.
Nervi, P. L. (1955b). Concrete and structural form. *The Architect and Building News, 208*(27), 523–529.
Utzon, J. (2000). *Sydney opera house – Utzon design principles*. The Vision: 1.1 (p. 22). The Sydney Opera House, Utzon Document.
Utzon, J. (2002). The innermost being of architecture (1948). In Weston, R. (Ed.), *Utzon inspiration vision architecture*. Copenhagen: Hellerup Edition Bløndal.

Online

Herron, L. (2015). Accessed September 12, 2016, from http://greenbuildingelements.com/2015/03/25/sustainability-key-to-sydney-opera-house

Pritzker Architecture Prize. (2016). Accessed December 13, 2016, from http://ecoworldly.com/2015/01/02/the-sydney-opera-house

References 3.9

Arslan, S., & Sorguc, A G. (2004). Similarities between "structures in nature"and "man-made structures": Biomimesis in architecture. In Collins, M. W., & Brebbia, C. A. (Eds.), *Design and Nature* (p. 48, Vol. II). Department of Architecture, Middle East Technical University, Turkey. WIT Press.

Carter, A., Kirkegaard, P. H., & Tyrell, R. (2013b). The nature of tectonic architecture and structural design. In P. J. da Sousa Cruz (Ed.), *Structures and architecture: New concepts, applications and challenges* (p. 237). London: Taylor and Francis Group.

Palumbo, P. (2015). Chair of the Jury of the Pritzker Architecture Prize.

Khan, F. R. (1980b). Structural aesthetics in architecture and its social and technological relevance. In *Zeitschrift: IABSE congress report AIPC=IVBH, Kongressbericht Nov* (p. 145).

Mentegazzi, E. (2014). In Mentegazzi, E. M. A. (Ed.), *Parametricism: The act of change in architecture* (pp. 17, 38). Department of the Build Enviroment. Eindhoven: AnArchi.

Schumacher, P. (2016a). *Parametricism – The Parametric Paradigm and the Formation of a New Style* (p. 619).

Online

Glancey, J. (2004). Accessed October 12, 2016, from http://www.theguardian.com/artanddesign/2004/oct/04/architecture

Pritzker Architecture Prize. (2015). Accessed October 11, 2016, from http://www.archdaily.com/607935/frei-otto-named-2015-pritzker-laureate

Palumbo, P. (2016). Chair: Pritzker Architecture Prize 2015 Jury. Accessed October 15, 2016, from http://www.pritzkerprize.com/2015/tributes-frei-otto

Hadid, Z. (2016). Accessed October 6, 2016, from http://www.pritzkerprize.com/2015/tributes-frei-otto

Gehry, F. (2015. Accessed October 6, 2016, from http://www.pritzkerprize.com/2015/tributes-frei-otto

Schumacher, P. (2016b). Accessed October 11, 2016, from https://en.wikipedia.org/wiki/Parametricism

References 3.10

Groys, B. (2013). Becoming revolutionary: On Kazimir Malevich 6, 8, 9. In: *Booklet: Russian Red (Wasiliew J.). 21.09 to 25.10.2013.*

Milne, R. J. W. (1997b). *Structural engineering: History and development* (pp. 42, 44, 46, 47, 48, 49). London/New York: CRC Press/T&G Group.

Sudjic, D. (2014). Director of the Design Museum; Prof. Milner, J.: Courtauld Institute of Art; Borchard-Hume, A.: Curator: Malevich Tate Museum. In Documentary: Hall, M. Producer: *Secret Knowledge*. Accessed on: BBC World TV Arts and Culture on 10.04.2016.

Online

Duncan, J. (2016). http://www.dezeen.com/2015/09/24/zaha-hadid-royal-gold-medal-architecture-2016-riba/
Groys, B. (2016). Accessed November 6, 2016, from http://www.e-flux.com/journal/becoming-revolutionary-on-kazimir-malevich
Hadid, Z. (2013). Heydar Aliyev Center/Zaha Hadid Architects, 14 Nov 2013. ArchDaily. Accessed November 5, 2016, from www.archdaily.com/448774/heydar-aliyev-center-zaha-hadid-architects
Marcus, F. (2014). Accessed November 5, 2016, from http://www.dezeen.com/2014/07/02/design-museum-designs-of-the-year-prize-zaha-hadid-heydar-aliyev-centre-azerba
Mark, L. (2014). Accessed November 7, 2016, from https://www.architectsjournal.co.uk/news/hadid-triumphs-in-design-of-the-year-awards/8661743.article
Parametricism. (2016). Accessed October 11, 2016, from https://en.wikipedia.org/wiki/Parametricism

Chapter 4
Comparative Structural Methods

Abstract Over a span of over 550 years, widely diverging operational methods employed by three architects in particular are brought into sharper focus; those of Filippo Brunelleschi, Antoni Gaudi and Pier Luigi Nervi. In this chapter, it is noted that they share common elemental factors that influenced their work and determined their design philosophies and structural principles, however, their methods of execution differed: Brunelleschi (Santa Maria del Fiore, Florence) embarked on a gravity defying modus to construct the dome over the transept of the cathedral without scaffolding; Gaudi (La Sagrada Familia, Barcelona) employed models involving ropes and sacks of lead shot to achieve his Nature-inspired soaring vaults and spires; and Nervi (Turin Exhibition Hall C) believed that precasting, the use of ferro-cement, and methods of roof construction with 'travelling' formwork, was an important step in construction that could deliver a finished state and define the aesthetic of a structure.

4.1 Santa Maria del Fiore, Florence (Il Duomo) Filippo Brunelleschi

Florence began to build a new cathedral in 1296. Wars, politics and plague slowed construction, so work on the dome didn't commence for more than a century. The structure and methods of construction of the dome for the cathedral of Santa Maria in Florence have roused considerable interest down the centuries.

Architects had to compete to design each stage of the project. The use of scale models was common in the planning of large structures and goes back to antiquity. Brunelleschi submitted a model to the Opera commission to convey both his concept and its practicality, using the model as a test bed to validate his methods of construction. Made of wood, brick, and mortar, it was large enough to allow members of the commission to walk inside and inspect its interior. The dome soars to a height of nearly 90 m with its vertical cross sections of the dome that are pointed Gothic arches, a wonder of ingenuity and achievement that outshone any other structure in the early Renaissance Italy (Hahn 2012a).

"In overcoming the problem of constructing the dome, Brunelleschi's solution is revealing about the ingenuity of the clockmaker and tinker himself" (Mueller and Yoder 2014). When his army of masons and stone-cutters, carpenters, blacksmiths,

D. Thomas, *Masters of the Structural Aesthetic*, DOI 10.1007/978-981-10-5445-7_4

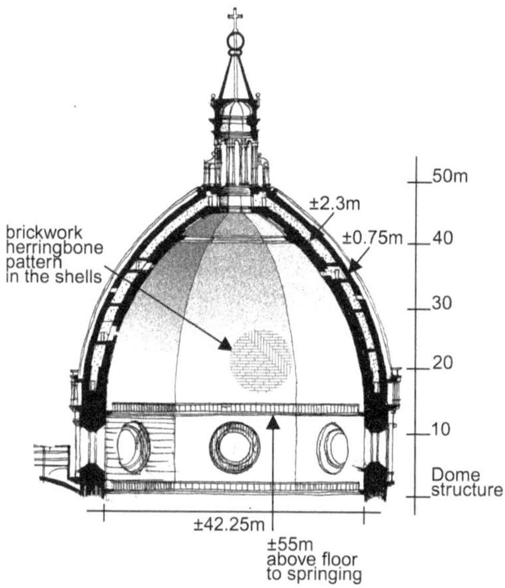

Fig. 4.1 Santa Maria del Fiore, Florence: The Dome, 1420–1436. For about the first 6 m the two shells of the dome were built with ashlar—big, cut, rectangular blocks of sandstone, followed by an interlocking and spiraling herringbone pattern for brickwork of differing shapes—rectangular, triangular and bricks with flanges (Baptista 1970). The shells were further reinforced by encircling sandstone cross-ties at four levels, from the springing to the apex. These were fashioned from 2.75 m sandstone blocks held together with iron clamps. In addition to these stone cross-ties, there wooden chains made from chestnut beams, strengthened by oak segments and attached to each other with iron pins

lead beaters, barrel-makers, water carriers, and other craftsmen were puzzled by construction details, it is said that he'd shape a model out of wax or clay or a carving to demonstrate what he wanted (Fig. 4.1).

The working crew was organised into eight teams so that the eight sections of the two shells could grow simultaneously, one octagonal ring at a time. During the construction of a ring the brick and mortar structure needed to be supported to prevent it from falling inward. However, a completed ring was stable on its own, compressed into place by the surrounding shell—formwork would therefore have been unnecessary. In the execution, work was carried out from platforms so that working in teams of eight the crews could progress consistently and simultaneously to ensure that the completed work would intrinsically remain stable (Hahn 2012a). For the roofing, terra-cotta roof tiles and interior patterning conceal many of the construction details.

As master builder Brunelleschi showed his human side, even ordering that the workers' wine be diluted with water. He took care of the safety of his army of masons, stonecutters, carpenters, blacksmiths, lead beaters, barrelmakers, water carriers and other craftsmen, by adding balustrades to suspended platforms to prevent accidents by falling.

Brunelleschi's methods of construction were ingenious. By using counter-weighted cranes, and other devices designed by him, he was able to move bricks, stones and mortar with ease and precision, even at the base for the lantern. Mueller and Yoder (2014) describes the construction methods, where Brunelleschi devised a three-speed hoist powered by oxen to turn the tiller and invented a system to power the gears, pulleys, screws and driveshafts to hoist heavy materials to the required levels. By using a special heavy rope weighing 454 kg, and over 180 m long, threaded through a clutch, the system could be reversed without having to turn the oxen around. A *castello* (tall crane) 20 m high with counterweights could move loads laterally. Sketches of the crane were depicted by Leonardo who resided in the nearby Tuscany town of Vinci, and confirm the details of Brunelleschi's invention.

The completion of the dome placed Florence ahead of other rival cities. At the time, many believed it was too big and ambitious to build, but architect Filippo Brunelleschi had succeeded using methods that experts don't fully understand nearly six centuries after its construction.

For 40 years Florentine architect Massimo Ricci, using a method of trial and error and scale models of Brunelleschi's innovative equipment, has scoured the interior and exterior of the dome for clues, mapping and cross-referencing against archival document, visible clues concerning the dome's construction (Ricci 2014). In a park on the south bank of the Arno River and about 1 km downstream from the cathedral, since 1989 Ricci has been building a model of the dome to a proportional scale of about 1:25. About half a million bricks have been used to form the base. Although useful to determine the geometry, the structuring method Brunelleschi used for the bonding of bricks is of equal importance. However, in the process of building the model, Ricci (2014) may have solved one of Brunelleschi's 'biggest secrets'; the manner in which the web of fixed and mobile chains was used to position each brick, beam and block so that the eight sides of the dome would arc towards the centre at the same angle.

Documentary references to 'the star of the cupola' led Ricci to experiment by suspending a star-shaped hub in the centre of his model of the dome. By stretching eight radiating chains outwards and downwards to the corners of the octagonal plan of his model, and linking these eight chains with horizontal ropes that trace the arcs of the eight sides of the octagon. Seen from above, these ropes resemble the petals of a flower (Ricci 2014).

4.2 La Sagrada Familia, Barcelona: Antoni Gaudi

The La Sagrada Familia is still under construction. The only existing copy of Gaudí's last recorded blue prints was destroyed in 1938 during the Spanish Civil War, and this has made it very difficult for his workers to complete the cathedral in

Fig. 4.2 La Sagrada Familias, Barcelona: Antoni Gaudi. *Right*: Under construction (circa 1990). *Below left*: model of the vaulting. *Below right*: interior vaulting present day. Gaudí studied and developed a new method of structural calculation based on models involving ropes and small sacks of lead shot. The playful and surreal soaring vaults and spires relied on Gaudi's careful and elaborate studies of complex systems of loaded strings

the fashion Gaudí most likely would have wished. Differences between Gaudí's original design and the new work are discernible, but on the whole are consistent with his own distinct, irregular and fantastically sculptural style (Fig. 4.2).

The hallmark of Antoni Gaudi's structural principles is that he transcended the conventional, by responding to the spirit of the age of Futurism and sourcing his inspiration from Nature. In cross-section, for example, his columns in their simplest form took a polygon that "by rotation gradually increased its number of sides, passing slowly to an almost circular section, and eventually to a circle" (Lorenzi and Mauro 2010). This cross-sectional design theme became more complex by duplication so that one column rotated clockwise and the other counter-clockwise. By extension of this principle, a progression that commences with a square cross-section, transforms into an octagon, then a 16-sided polygon. The columns of the La Sagrada Familia follow this theme ranging from four to twelve initial sides that increase in number from bottom to top. Intermediate columns in turn form the starting point for smaller columns that split and spread the forces of the vaults.

"The outline of the church was traced on wood and placed on a ceiling, with ropes that hung from the points where columns had to be placed. Antoni Gaudí made extensive use of catenary shapes not only in the Sagrada Família but in most of his architectural work. A 'catenary' (derived from the Latin word 'catena' meaning 'chain') is a curve U-like in shape and similar to a parabola that represents the shape it assumes—idealised in a hanging chain—when it is supported at its fixed ends and acted upon by gravity. Sacks of pellets—weighing just a fraction of the weight the real arches would have to support—were hung from each arch formed by the ropes. After taking photographs of the resulting model, shot from various angles, turning then upside-down the lines of tension formed by the ropes and weights would eventually indicate the pressure lines of the structure envisaged" (Lorenzi and Mauro 2010). The catenary is very important in architecture, in fact it is the ideal curve for an arch that supports not only its own weight.

As an approximation, when the centre line of an arch follows the curve of an inverted catenary, then the arch is known to endure only pure compression so that no significant torsional moments occur inside the material and no specific buttress is required, since the forces acting on the arch at the two endpoints are tangent to its centreline. There can be no better demonstration of how Gaudi's modelling method for the structure of the cathedral correlate aesthetically to forms that are found in Nature.

Some of the sandstone that Gaudi used is now deteriorating so that maintenance and construction go on in parallel, while synthetic sandstone with the same appearance of the original is being used for the new construction work. The architectural work of Gaudí is remarkable for its range of forms and textures, using the geometries that he devised that are complex as well as free and expressive (Hahn 2012b).

4.3 Palazzetto della Sport, Rome: Pier Luigi Nervi

Pier Luigi Nervi was responsible for the design of his structures both for their construction as engineer and director in the firm that built them, two distinct functions that contributed largely to his success. The opportunity of uniting the two aspects of design and execution, in contemporary terms, have tended to become separated into two distinct functions.

The Palazzetto della Sport is covered with a 61 m diameter dome over a floor area of 4790 m^2, and carried on 36 external supports seach a low raking Y continuing the line of the roof; a second vertical ring, tapering down from the fork of the Y makes these in situ concrete supports self-supporting (Fig. 4.3).

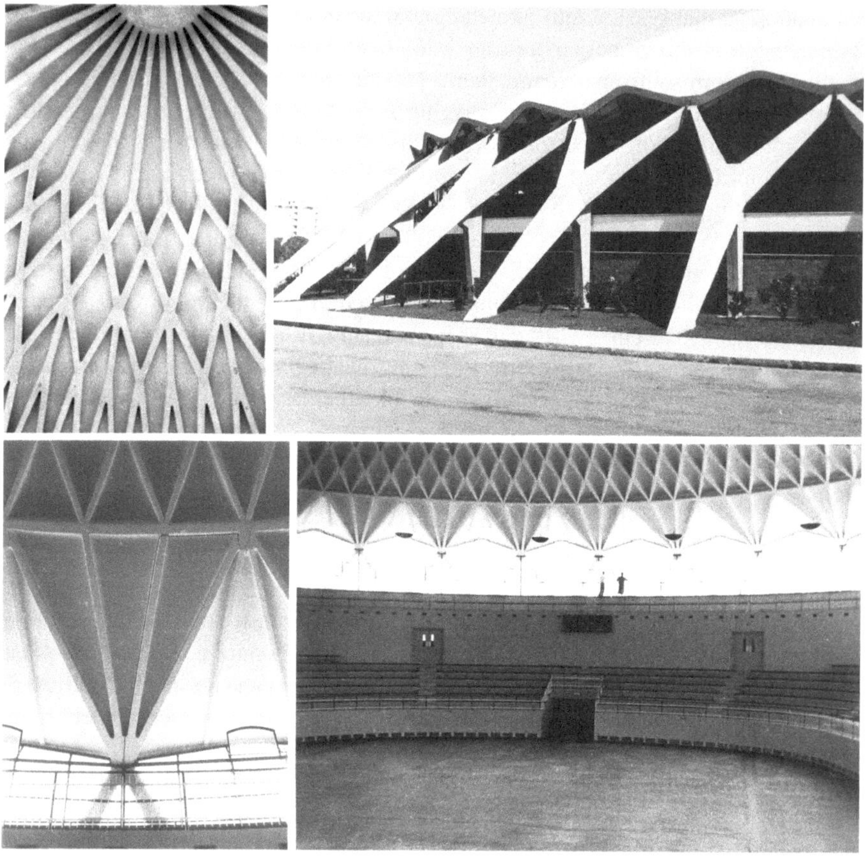

Fig. 4.3 Palazzetto della Sport, Rome: Pier Luigi Nervi. The emphasis was placed on the full development of reinforced concrete, not only from the engineering, but also from the architectural point of view, a clear liberation from the restrictions imposed by timber formwork, which in fact obliges the designer to conform to the pattern of timber construction. This was in direct opposition to Nervi's belief that precasting, the use of ferro-cement, and methods of roof construction with 'travelling' ferro-cement formwork, was an important step in demonstrating his approach

Nervi (1955) observed that "The present moment in architecture is full of promise, but the danger of slipping into structural formalism should never be overlooked: alarming symptoms of it can already be seen in the architectural work illustrated in periodicals of all countries. This derives from the fact that too often, through lack of understanding of its structural and constructional essence, a structure is considered solely on the basis of its external appearance, which people try to adapt to a variety of different problems both as regards dimensions and strength. The result is always unfortunate. I am absolutely certain that the prime condition of architectural expression in a structure is the correctness, I might say, the inevitability of structural design."

In Italy at the time of Nervi's productive years, a system known as 'competition—tender' was being used increasingly. The method consisted of inviting a member of a firm known to be qualified from a technical point of view to submit a tender including the actual design, as well as a price quotation. The required design was to be based on the outline provided by the commissioning authority and these outlines always allowed ample freedom for the best architectural and structural solutions to the problem. The commissioning authority made separate examinations of the technical and economic data and selected the design which seemed the most satisfactory from all points of view.

The design that is good from both the aesthetic and technical points of view is generally sufficiently economical. In any case, the commissioning authority had full data on which to base a choice and could even accept a tender which was not the lowest, if the qualities of the design were such as to outweigh a small difference in cost. In some cases the tender called for a 'lump sum contract', so that, except for unforeseen circumstances, the commissioning authority knew exactly how much the finished work will cost.

The advantages of such a system are obvious: competition stimulates the designer-contractor to develop the most suitable design from an economical point of view, and study new and more efficient methods for its execution. The extensive knowledge of materials, which Nervi possessed as a practical contractor, his realisation of the limitations and difficulties of the actual execution and his study of the means of overcoming them, ensured a realistic approach to the design, but also suggested daring yet rational solutions that may have seemed impossible to the purely professional designer. Preliminary tests and experiments, even quite costly ones, could be carried out by the designer-contractor within his own organisation, whereas this would be impractical for the purely professional practitioner.

It is clear that these observations all point to the fundamental importance of structurally honest solutions in the design of reinforced concrete structures. Arguably, none of Nervi's buildings would have been successful, and in many ways possible, if the method of construction as an integral part of the design had not been studied from beginning to end (Thomas 1958).

Ferro-cement by virtue of its independence of formwork, by its intrinsic lightness (the units are less than 4 cm thick) and its strength, has provided the simplest and most satisfactory solution to the complex problem.

Without the constructional qualities of ferro-cement the entire architectural-structural conception would have been different proving Nervi's point when he emphasises the close ties between the architectural conception and construction reality, where limited time is available for the work. For the advantage of lightness for the large span of the main structure, it was essential to find a solution which would unite the possibility of prefabrication to low weight, good appearance and good thermal and sound insulation.

For the perimeter roof of the Palazzetto Nervi evolved the corrugated beams 2 cm thick (for a 9 m span) and owing their strength to their form, easily precast, easily handled and which placed side by side and joined together structurally by a thin slab, formed a light roof of pleasant appearance and with good sound and thermal insulation.

The thrust of the dome through the supports is finally restrained by a circular prestressed tension ring, 2.5 m wide and 81.5 m in diameter. Based on the Freyssinet System, the prestressing is by means of two layers of seven cables, overlapping and anchoring at the foot of each support. It was constructed on a 10.16 mm thick bed of lean concrete covered with 5.08 mm of sand and a layer of bitumenised paper. This beam distributes the load to the ground.

The supports were cast against a carefully designed timber framework and left as they came from the forms; the combination of good concrete with the strong timber markings give them a look of taut strength. (Nervi, however, was not satisfied: the concrete could have been better). Besides transmitting the load of the dome to the foundations and its self-weight, the supports are designed to resist horizontal forces due to wind pressure.

The dome is made up of 1620 precast concrete units cast on site, on forms which were, in their turn, made on a template, which produced exactly a section of the dome. The division of the roof into precast elements is a function of its structure—a structure both simple, practical and of proved economy—but the actual shape of the units was, designed to suit an aesthetic objective: it is the interplay of the diamond-shaped sections of this roof, narrowing to the apex and broadening to the base, to end as triangles tapering to meet the converging supports, that give this dome is wonderful grace.

The units were being cast while the foundations and supports were being constructed, and work was progressive throughout; the screed reinforcement and surface screed itself were placed as soon as the dome units were in position and while placing of the remainder of the units was still in progress. Two cranes, one at the centre and worked around the periphery. The units were placed on a series of concentric rails fixed directly to the tubular steel scaffolding, and so arranged that the unit rested on the rails, top and bottom.

The dome was calculated as a membrane; on its completion and removal of the formwork and temporary supports the static behaviour of the structure was checked by means of strain gauges placed in the fork of the Y supports. These readings showed the perfect 'centering' of the loads and, confirmed the accuracy of the preliminary calculations.

Inside the building the tiers of seats are in situ concrete. The designers had wished to precast them and they do in fact lack the fine detailing they might otherwise have had. Together they form on plan two crescents, giving the arena an oval shape. The arena floor itself is 1 m. below ground level; from the arena to the roof apex the height is 21 m.

Construction of the building began on 26th July 1956 and was completed on 15th September 1957. Construction of the actual dome took 40 days, several of which were lost through bad weather.

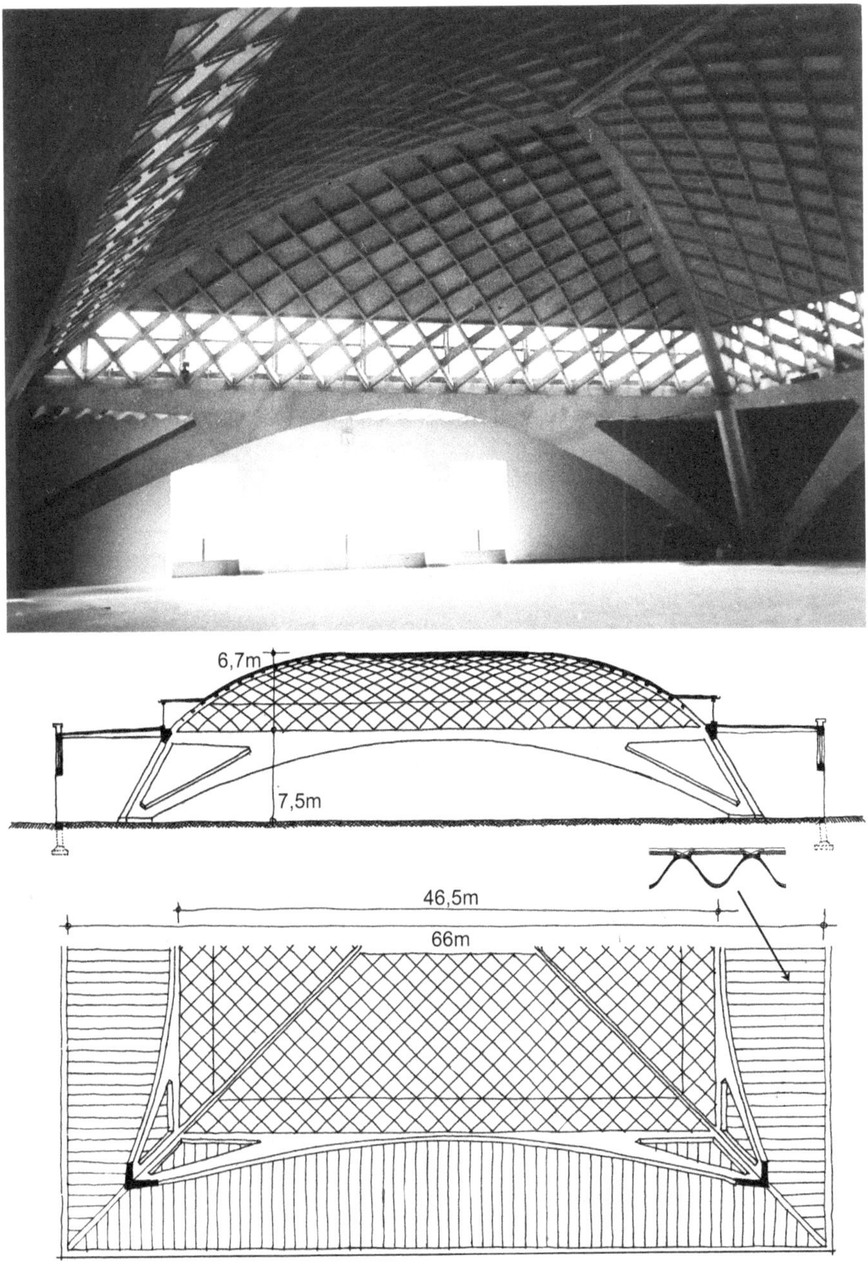

Fig. 4.4 Exhibition Hall C, Turin, Italy. A structure of this type built by the usual methods of construction would have required an amount of formwork at significant cost and time, while the use of conventional precast reinforced concrete units would have been equally difficult on account of their laborious manufacture and their excessive weight

4.3.1 Turin Exhibition Hall C

The structural principle for Exhibition Hall C differs from Hall B. For the vaulting over the central hall and the perimeter roofs of Hall C the juxtaposition of precast ferro-cement units was formed, joined by reinforced concrete ribs cased in situ in the troughs and at the crests of the corrugations.

For the surrounding flat-roofed portion, which spans 10 m Nervi designed a system of corrugated beams in ferro-cement, precast at the same time as the vault units. These beams are placed side by side and finished with a lightweight screed. Erection proved quick and easy. The beams in question are 19 mm thick at the top, increasing to 38 mm at the soffite to provide for the necessary reinforcement. They were cast in moulds which in turn had been made in a plaster mould. The visible underside of the beam, which was in contact with the forms during casting, is perfectly regular and smooth, with perfection of surface that could never be obtained by any usual finishing processes (Fig. 4.4).

4.3.2 Palace of Sport, Rome

This building provided Nervi with the opportunity to put into practice the use of ferro-cement units to form the dome, tapering towards a central compression band (Fig. 4.5).

Nervi is often accused of having mixed up the role of the engineer with that of the contractor, a point of view which regards the contractor as a purely functional tool. But just because of this constant mixing of the two roles both Perret and Nervi achieved that free mastery and complete control of the structure which was the foundation of their success. Both were exceptional exponents in the history of modern architecture. Through them the art and wisdom of architecture, which were the strength of the old master-builders, are carried on (Rogers 1957).

4.3.3 Stadium Flaminio, Rome

From all outward appearances the Stadium is just another stadium. However, Nervi achieved something of finesse in the very pleasing lines of this 1960 Olympic arena. Almost the entire structure is prefabricated including the seats which are designed to span between inclined concrete beams.

For lightness the cantilevered corrugated roof over one section of seats is of precast ferro-cement units, each weighing 4 tons, erected into place by two cranes. Each unit has portholes for further lightness and is bonded into the in situ structure from which it cantilevers.

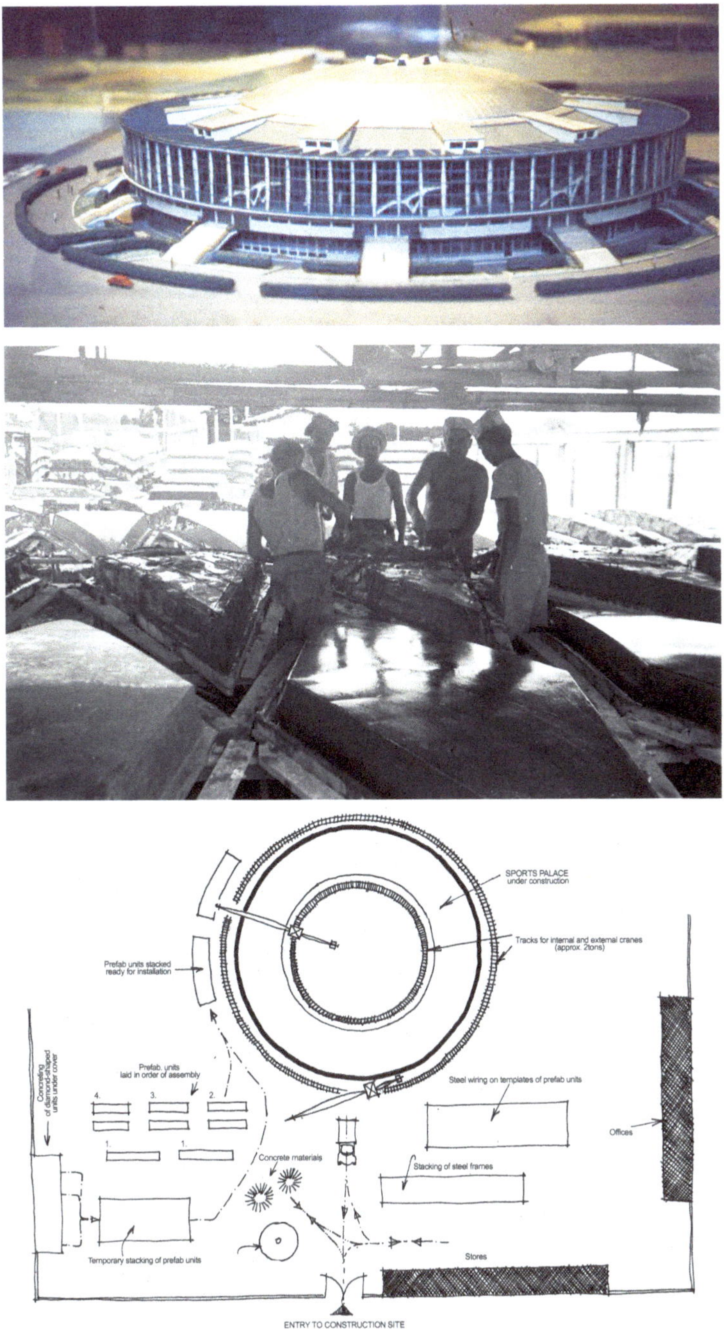

Fig. 4.5 Palace of Sport Rome. The building is a domical version of the segmented vaulting of the main Exhibition Hall C in Turin. The corrugated roof is connected to the main supporting columns by fan-shaped ferro-cement units springing from inclined supporting members. The builder's yard at the Palace of Sport. Typically, the forms were concrete made with a regular and satin smooth

Fig. 4.6 The 'Flaminio' Stadium. From all appearances just another stadium, however, Nervi achieved achieved finess in the pleasing lines of this Olympic arena. Almost the entire structure is prefabricated including the seating that spans between inclined concrete beams

The cost of the project was estimated to be £4,750,000 and seats 50,000 spectators (Fig. 4.6).

> In the last few years I have had the opportunity of collaborating with architects of great capacities and artistic sensibility, in the conception and development of several outstanding projects, and I must say that this collaboration has been carried out with mutual satisfaction and with results which I consider interesting. I would mention in this connection, the UNESCO building in Paris, in which I collaborated with Breuer and Zehrfuss, the Pirelli offices in Milan, the project for the Central Station in Naples.
> Nervi (1955)

The question arises as to how to gain insight into the spirit of Nervi's own working method to further current design thinking with contemporary

---

Fig. 4.5 (continued) finish which gives the perfect surface to the soffite of the box-shaped units. These are 25 mm thick, and are made of ferro-cement. The concept of this material is based on the elementary and familiar observation that the elasticity of a reinforced concrete member increases in proportion to the subdivision and distribution of the reinforcement throughout the mass. The richness of the mortar (800 kg of best quality cement to 1 m^3 of sand) allowed the units to be demoulded in either 2 or 3 days, according to the outside temperature. Lifting and placing of the units proceeded regularly and enable about 300 m^2 of roof to be completed each day. The construction would normally take place in three stages to get the fullest possible use from the movable formwork computational tools and knowledge of pattern formation in Nature, and the significant advances in constructional technology.

References

References 4.1

Baptista, F. G., Twombly, M., & Snodgrass, E. (1970). National Geographic Magazine Staff; Gebreyesus, F., Margaret, N. G., Picture text; Williams, A. R., NGM Staff. Drawing based on a fresco by Masaccio (CA141426), Brancacci Chapel, Florence model image: Museo Dell-Opera del Duomo, Florence, Ital/Bridgeman Art Library. Sources: Mainstone, R.; della Negra, R., University of Ferrara, Italy; Ricci, M., Forum Unesco – University and Heritage Polytechnic University of Valencia, Spain; Gurrieri, F., University of Florence. In: Infograph, National Geographic.

Hahn, A. (2012a). *Mathematical excursions to the world's great buildings* (p. 119, 121, 122). Princeton: Princeton University Press.

Online

Mueller, T., & Yoder, D. (2014). *Il Duomo. Brunelleschi's dome. How did a hot-tempered goldsmith with no formal architectural training create the most miraculous edifice of the renaissance?* http://ngm.nationalgeographic.com/2014/02/il-duomo/mueller-text

Ricci, M. (2014). Accessed November 26, 2016, from http://news.nationalgeographic.com/news/2014/02/140210-duomo-florence-brunelleschi-cathedral-architecture/

References 4.2

Hahn, A. (2012b). *Mathematical excursions to the world's great buildings* (pp. 259–261). Princeton: Princeton University Press.

Lorenzi, M. G., & Mauro, F. (2010). Art and mathematics in Antoni Gaudi's architecture: "La Sagrada Familia". *Journal of Applied Mathematics (Aplimat), 3*(1), 132–137.

References 4.3

Nervi, P. L. (1955). Concrete and structural form. *The Architect and Building News, 208*(27), 523–529.

Rogers, E. N. (1957). *The works of Pier Luigi Nervi*. vi, ix, x, xi (Preface by Pier Luigi Nervi)

Thomas, D. C. (1958, July/September). *A study of the works of Pier Luigi Nervi*. Report on research of the new structural techniques and potentialities of ferro-cement.

Afterword

Modern day individualism and eclectic trends in architecture, particularly in the developed world, have tended to resort to exhibitionistic solutions, choosing to give expression to experiential environments, even narcissism and, in some instances, buildings that are alienated from rational design. Invariably it can be the patrons who indulge in fanciful and extravagant prestige buildings and play a role by encouraging architects to 'strut their stuff'.

From prehistory to the present day, architecture has meant shelter and amenity for humankind to survive and to conduct life creatively, spiritually and with fulfilment. The main stem of architecture is a remarkable history of imagination, however, there is an indulgent imbalance when contemporary work is placed alongside past periods. Traditional vernacular architecture, mostly of pre-industrial times, invariably 'got it right' where, with relatively limited means, extraordinary creative enterprise produced architecture often of the most distinctive character and ingenuity in the creation of shelter. Manual labour and crafts, materials, and technical and technological expertise either constrained or furthered the ambitions of those who commissioned the work or were responsible for crafting it. Inspiring examples are found in both urban and rural situations around the world where functionality, shelter, security and status were expressed in the vernacular, suggesting again the potential for the aesthetic in architecture to be achieved through honesty without the concealment of cladding or the membrane of the ubiquitous glass curtain wall, and without the greater dependence on energy-consuming means for the comfort of the occupants. Despite the demonstrated time-saving, economical, and aesthetic capabilities of ferro-cement in construction, since Pier Luigi Nervi its potential in contemporary work has not been fully explored.

Computer-aided technology and structural innovations have encouraged a new value system often in contrast to how honest architecture is made. The challenge for the future lies within the power of architects and structural engineers to work in unison with digital technologists, resource economists and other relevant disciplines to write holistic briefs for their work, for the return to honest architecture and indeed to exploit the potential of the structural aesthetic meaningfully.

D. Thomas, *Masters of the Structural Aesthetic*, DOI 10.1007/978-981-10-5445-7

> Meanwhile, until such time as the training of the complete architect can be achieved, good results can be obtained through the sincere collaboration of many people, each contributing the specific knowledge lacked by others. Architect, Engineer, Constructor can in this way bring about that union of art and science that is necessary to the solution of any constructional problem. To deal with the even more ambitious architectural projects of the near future, the architect must possess—and synthesise in himself—aesthetic sensibility, profound understanding of structural needs, and a precise knowledge of methods, possibilities and limitations of constructional techniques.
> (Nervi 1955)

In today's world, with unprecedented population growth and the unsatiable demand on resources, there is shrinking space for extravagant architectural design. With the challenge of sustainability haunting modern life, architects should be reflecting on the historical legacies of the past, in order to understand the essence as to why certain architectural typologies have endured and have past the test of sustainable building design. Roman and Gothic structures stand out in terms of both their durability as well as their sustainable functionality. Modern structures have a hard act to follow, considering the avenues of individuality and indulgences that have become the hallmark across a wide spectrum globally.

In considering 'fanciful' trends in the Western world, as well as in terms of global architectural evolution as a whole, Sir Banister Fletcher, an accepted authority on the history of architecture, postulated that the nineteenth century was an era of revivals of past styles but that individual architects 'are unlikely to return to any one systematised style'. Nevertheless, 'architecture still continues to reflect the thought of the day, the needs and aspirations of the people, and is an index of the social forces at work...' (Fletcher 1948). Added to the 'social forces', in the 21st Century, technological advances in all spheres will irrevocably augment the purely social influence on architectural design.

A greater reliance on the predominance of the structural aesthetic, rather than on embellishment or exhibitionism purely for experiential pleasure, could be key where architecture becomes an honest response to the determinants outlined in the chapters of this book. A twenty-first century paradigm that is capable of building on the durable architecture of past eras suggests mitigation based on the constructs of recover, recycle, re-use and re-purpose that could contribute to longevity in architecture and be in keeping with the ethic of global responsibility.

References

Fletcher, B. (1948). The historical styles. Introduction 1. In *A history of architecture on the comparative method* (14th ed., p. 9). London: B.T. Batsford Ltd.

Nervi, P. L. (1955, October 14). *Notes from a lecture to organised by the Institution of Structural Engineers and Joint Committee on Structural Concrete at Friends' Meeting House*, Euston Road, London.

Notes: On Ferro-cement: Its Characteristics and Potentialities

The manufacture of thin slabs of cement mortar reinforced with super-imposed layers of wire mesh and small diameter bars gives a product with a high degree of elasticity and resistance to cracking, and which does not require formwork. Nervi carried out experiments close on 17 years; he achieved results, found potentialities, proving the discovery of a material he called 'ferro-cemento'

The concept of this material is based on the elementary and familiar observation that the elasticity of a reinforced concrete member increases in proportion to the subdivision and distribution of the reinforcement throughout the mass. The richness of the mortar (800 kg of best quality cement to 1 m^3 of sand) allowed the units to be demoulded in either 2 or 3 days, according to the outside temperature. Lifting and placing of the units proceeded regularly and enable about 300 m^2 of roof to be completed each day. The construction would normally take place in three stages to get the fullest possible use from the movable formwork (Fig. A.1).

The mortar is made with good quality sand, in proportions varying from 800–1000 kg of cement per cubic metre of sand. In this way very thin slabs were made which were exceptionally flexible and elastic and extremely strong.

To increase the thickness and strength of the slabs without using more than 10–12 layers of mesh, Nervi tried introducing between the layers of mesh on or more layers of bars of a diameter 6 mm, without losing the qualities of the material formed with the mesh alone.

In every case the thickness of the finished prefabricated unit was only a very little greater than that of the assembled layers mesh, the difference bing only a much as was necessary to provide enough cover for the steel.

The material obtained has very little in common with normal reinforced concrete, possessing the mechanical characteristics of completely homogeneous material.

Its most important qualities are the great elasticity and resistance to cracking given to the cement mortar by the extreme subdivision and distribution of the reinforcement, and the fact that the mortar itself can be applied without the need for formwork and remains held perfectly in place by the mesh.

D. Thomas, *Masters of the Structural Aesthetic*, DOI 10.1007/978-981-10-5445-7

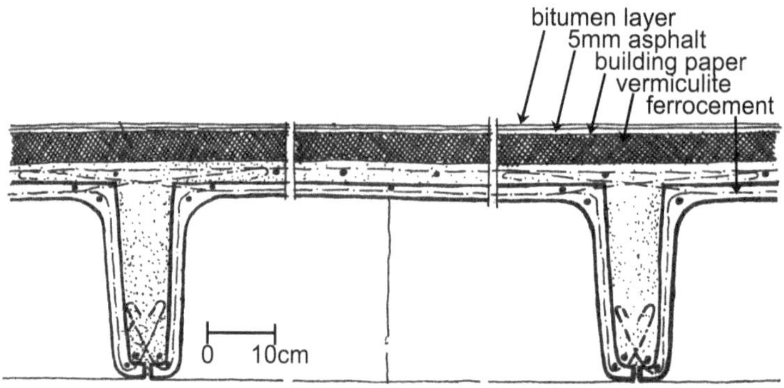

Fig. A.1 Ferro-cement units. Detail cross-section of the typical ferro-cement units with finishing layers. The reinforcement in some cases consisted of 5 mm wire arranged in both directions and a double layer of mesh weighing just over 0.45 kg per 0.85 m^2. The top-most rings of the units have additional reinforcement in the form of 15.2 and 23 mm bars. The reinforcing wires project from the sides of the units and assist in uniting them structurally with the in situ concrete placed in channels between them and as a screed over the top. This in situ screed is further reinforced with 7.6 mm steel placed concentrically and radially. A plastic film sprayed onto the surface provides the waterproofing

These characteristics opened up a number of interesting fields of application, among them naval work for building small ships.

In 1943 more accurate experiments were carried out, some under the direction of Professor Martinelli. These experiments were aimed at drawing up data concerning the mechanical characteristics, the resistance to shock and the extensibility of slabs and samples of ferro-cement of different thicknesses and with different percentages of reinforcement.

The shock resistance tests revealed not only the high strength of slabs of only 3 cm thick (a weight of 250 kg was dropped on to a sample 1.5 m^2, from different heights increasing approximately to 3 m), but also the fact that failure, when it occurred, did not consist in an actual hole in the structure, but rather a weakening of the wire mesh, a relatively dispersed breaking away of the mortar, the pieces of which still retained a certain cohesion, and a certain resistance to the passage of water.

Several completely successful adaptations of ferro-cement to naval craft construction have been achieved. While these were being undertaken, in 1947 Prof Oberti started a new series of more accurate tests in the Polytechnic Laboratory in Milan, with the object of providing numerical data on the degree of extension without visible cracking obtainable with specimens of ferro-cement.

These tests show the vital importance of the percentage of steel in relation to the concrete mass in obtaining maximum extension. With the layers of mesh giving a weight of steel of less than 100–200 kg to the cubic metre the extensibility remained approximately the same as unreinforced mortar. With weights of 400–500 kg there is a rapid increase in the extensibility, which went up five and more times the previous case.

It is noteworthy that a weight of steel of about 500 kg to the cubic metre is obtained when the slab is only a few millimetres (just sufficient to cover the steel) thicker than the layers of mesh themselves, placed in contact with one another. This is the natural proportion of the method.

Lower proportions may be obtained by increasing the thickness of the slab in relation to that of the steel, or by placing the layers of mesh further apart by means of spacing bars. The mechanical and constructional qualities become proportionally less as the steel proportion decreases.

The outcome depends on the desired extensibility, a definite preponderance of the metal reinforcement over the cement mortar, and its all over distribution.

Index

D. Thomas, *Masters of the Structural Aesthetic*, DOI 10.1007/978-981-10-5445-7

FSC
www.fsc.org
MIX
Papier | Fördert
gute Waldnutzung
FSC® C083411

Kolibri 360 GmbH
Ferdinand-Jühlke-Straße 7
99095 Erfurt, Deutschland
produktsicherheit@kolibri360.de